JOH[illegible]CKMAN

INTELLIGENT THOUGHT

John Brockman, editor of many books including *The Next Fifty Years, Curious Minds, What We Believe But Cannot Prove, My Einstein,* is also the author of *By the Late John Brockman* and *The Third Culture*. He is the founder and CEO of Brockman, Inc., a literary and software agency, and the publisher of the Web site *Edge*, the forum for leading scientists and thinkers to share their research with the general public. He lives in New York City.

www.edge.org

BOOKS BY JOHN BROCKMAN

AS AUTHOR:

By the Late John Brockman
37
Afterwords
The Third Culture: Beyond the Scientific Revolution
Digerati

AS EDITOR:

About Bateson
Speculations
Doing Science
Ways of Knowing
Creativity
The Greatest Inventions of the Past 2,000 Years
The Next Fifty Years: Science in the First Half of the Twenty-first Century
Curious Minds: How a Child Becomes a Scientist
The New Humanists: Science at the Edge
What We Believe But Cannot Prove: Today's Leading Thinkers on Science in the Age of Certainty
My Einstein: Essays by the World's Leading Thinkers on the Man, His Work, and His Legacy

AS COEDITOR:

How Things Are

INTELLIGENT THOUGHT

SCIENCE VERSUS THE INTELLIGENT DESIGN MOVEMENT

EDITED BY

JOHN BROCKMAN

VINTAGE BOOKS

A Division of Random House, Inc.

New York

FIRST VINTAGE BOOKS EDITION, MAY 2006

 Published in the United States by Vintage Books, a division of Random House, Inc., New York, and in Canada by Random House of Canada Limited, Toronto.

The Cataloging-in-Publication Data is on file at Library of Congress.

Vintage ISBN-10: 0-307-27722-4
Vintage ISBN-13: 978-0-307-27722-0

www.vintagebooks.com

Printed in the United States of America
10 9 8 7 6 5 4 3 2 1

Acknowledgments

I AM GRATEFUL to Marty Asher of Vintage Books for his continuing interest in science-minded ideas and his passionate support of this project. I am also indebted to my agent, Max Brockman, who immediately understood the need for this book. And as always, I am indebted to Sara Lippincott for her thoughtful and meticulous editing.

Contents

JOHN BROCKMAN

Introduction

SOMETHING RADICALLY NEW is in the air: new ways of understanding physical systems, new ways of thinking about thinking that call into question many of our basic assumptions. A realistic biology of the mind, advances in evolutionary biology, physics, information technology, genetics, neurobiology, psychology, engineering, the chemistry of materials—all are issues of critical importance with respect to what it means to be human. For the first time in our collective history, we have the tools and the will to undertake the scientific study of human nature.

"Nobody ever voted for the telephone, the automobile, for printing, for television, or for electricity," I wrote in 1969. And science-based reality continues: Nobody voted for the computational idea, for cybernetics, for the mathematical theory of information, for the Internet. Governments and politicians, operating through rearview mirrors, can only play catch-up. Science is the big news. Science is the important story. Science is public culture.

Yet at the same time, religious fundamentalism is on the rise around the world, and our own virulent domestic version of it, under the rubric of "intelligent design," by elbowing its way into the classroom abrogates the divide between church and state that has served this country so well for so long. Moreover, the intelligent-design (ID) movement imperils American global dominance in science and in so doing presents the gravest of threats to the American economy, which is driven by advances in science and in the technology derived therefrom.

This book—sixteen essays by leading scientists from several disciplines—is a thoughtful response to the bizarre claims made by the ID movement's advocates, whose only interest in science appears to be to replace it with beliefs consistent with those of the Middle Ages. School districts across the country—most notably in Kansas and later in Pennsylvania, where the anti-evolutionist tide was turned but undoubtedly not stopped—have been besieged by demands to "teach the debate," to "present the controversy," when, in actuality, there is no debate, no controversy. What there is, quite simply, is a duplicitous public-relations campaign funded by Christian fundamentalist interests.

In some ways, the media chatter provoked by the intelligent-design movement has made collective fools of large segments of the American public. Educated Americans are dumbstruck by the attempt of the state of Kansas to officially redefine science to include the supernatural. Europeans cannot believe that such an argument should be raging in the twenty-first century—and in the United States, of all places, the seat of our most advanced technology and a leader in so many areas of scientific research.

Fortunately, Judge John E. Jones III, of the United States District Court for the Middle District of Pennsylvania, in his December 2005 decision in *Kitzmiller et al. v. Dover Area School District*, has—temporarily, at least—halted ID in its tracks. Readers can rely on the writings of the scientists in this volume. They can also rely on the words of this lifelong Republican jurist, who was appointed to the court by the current President of the United States (himself a born-again Christian fundamentalist).

In his decision, Judge Jones writes:

> After a searching review of the record and applicable case law, we find that while ID [intelligent design] arguments may be true, a proposition on which the Court takes no position, ID is not science. We find that ID fails on three different levels, any one of which is sufficient to preclude a determination that ID is science. They are: (1) ID violates the centuries-old ground rules of science by invoking and permitting supernatural causation; (2) the argument of irreducible complexity, central to ID, employs the same flawed and illogical contrived dualism that doomed creation science in the 1980s; and (3) ID's negative attacks on evolution have been refuted by the scientific community . . . [I]t is additionally important to note that ID has failed to gain acceptance in the scientific community, it has not generated peer-reviewed publications, nor has it been the subject of testing and research.

In characterizing the Dover School Board's decision to offer intelligent design in the classroom as an alternative to the Darwinian theory of evolution by natural selection, he employs two words that will go down in history: "breathtaking inanity." To Judge Jones, it is "abundantly clear that the Board's ID Policy violates the Establishment Clause"—that is, the Constitution's First Amendment ("Congress shall make no law respecting an establishment of religion, or prohibiting the free exercise thereof"). He continues: "In making this determination, we have addressed the seminal question of whether ID is science. We have concluded that it is not, and moreover that ID cannot uncouple itself from its creationist, and thus religious, antecedents."

Given that all of science and reason is on the side of evolution, why bother with a book to make this case? If there is a positive side to the recent developments, it is that certain questions have been raised, and the forum thus created provides scientists with an unusual and broad public platform to explain how natural selection answers those questions. It is my firm belief that scientists should seize every opportunity to educate the general public. In this regard, I have asked the contributors—all of them prominent scientists and public intellectuals—to present essays on their personal view of the issues raised by the intelligent-design movement, and of the fact, power, and beauty of the Darwinian theory of evolution by natural selection.

Some of these essays deal with the bogus arguments put forward by the movement's proponents. What are the hallmarks of a genuine scientific theory? Can intelligent design truly be characterized as science? And if not, why not? Is ID, after all, religion in disguise? Is it, perhaps, a hoax? What are the dishonest tactics employed in promoting the ID movement, and how should these tactics be met?

Others are imaginative takes on Darwinian evolutionary theory: Among the topics addressed are recent discoveries in the fossil record, the nature of consciousness, the wellspring of ethics, directed panspermia (the notion that life on Earth originated elsewhere), "self-organization" in biology, the "intelligence" of the universe itself, and the natural selection of bio-friendly universes.

It is to be hoped that the ID movement, because of the very publicity that it has sought and achieved, will be seen by the majority of Americans for the giant step backward that it is. Our children are literally the future of our nation, which will increasingly need competent scientists and engineers to guide us through the coming technological revolutions—revolutions that are already under way all around us. There are examples in history of the collapse of great civilizations. There is no particular reason that the United States should be exempt from historical forces. The Visigoths are at the gates. Will we let them in?

INTELLIGENT THOUGHT

JERRY A. COYNE

Intelligent Design: The Faith That Dare Not Speak Its Name

Intelligent design is not an evangelic Christian thing, or a generally Christian thing or even a generically theistic thing. . . . Intelligent design is an emerging scientific research program. Design theorists attempt to demonstrate its merits fair and square in the scientific world—without appealing to religious authority.

—William A. Dembski,
The Design Revolution (2004)

[A]ny view of the sciences that leaves Christ out of the picture must be seen as fundamentally deficient. . . . [T]he conceptual soundness of a scientific theory cannot be maintained apart from Christ.

—William A. Dembski,
Intelligent Design: The Bridge Between Science & Theology (1999)

WELL, WHICH IS IT? Is intelligent design (ID) merely a sophisticated form of biblical creationism, as most biologists claim, or is it science—an alternative to Darwinism that deserves discussion in the science classroom? As the two quotations above imply, you won't find the answers in the writings of the leading advocates of ID.

The ambiguity is deliberate, for ID is a theory that must appeal to two distinct constituencies. To the secular public, ID proponents present their theory as pure science. This, after all, is their justification for a slick public-relations campaign promoting the teaching of ID in the public schools. But as is clear from the infamous "Wedge Document" of the Discovery Institute, a right-wing think tank in Seattle and the center for ID propaganda, intelligent design is part of a cunning effort to dethrone materialism from society and science and replace it with theism.[1] ID is simply biblical creationism updated and disguised to sneak evangelical Christianity past the First Amendment and open the classroom door to Jesus. The advocates of ID will admit this, but only to their second constituency, the sympathetic audience of evangelical Christians on whose support they rely.

Nevertheless, let us give the ID movement the benefit of

---

[1] The Wedge Document, an internal memorandum of the Institute, was leaked to the Internet in 1999 and later acknowledged by the Institute as authentic. It can be found at http://www.geocities.com/CapeCanaveral/Hangar/2437/wedge.html. It states: "The social consequences of materialism have been devastating. As symptoms, those consequences are certainly worth treating. However, we are convinced that in order to defeat materialism, we must cut it off at its source. That source is scientific materialism. This is precisely our strategy. If we view the predominant materialistic science as a giant tree, our strategy is intended to function as a 'wedge' that, while relatively small, can split the trunk when applied at its weakest points . . . Design theory promises to reverse the stifling dominance of the materialistic worldview, and to replace it with a science consonant with Christian and theistic convictions."

the doubt. Let us suppose that ID might indeed be an alternative and superior scientific theory—one that explains the natural world better than Darwinian evolution does. Can such an argument stand up to scrutiny? Is it time for Darwinian evolution to go the way of Newtonian mechanics, as a theory good for its time but ripe for replacement by a new paradigm? No. Not only is ID markedly inferior to Darwinism at explaining and understanding nature but in many ways it does not even fulfill the requirements of a scientific theory.

What are those requirements? A scientific theory isn't just a guess or speculation, it is a convincing explanatory framework for a body of evidence about the real world. A good scientific theory makes sense of wide-ranging data that were previously unexplained. In addition, a scientific theory must make testable predictions and be vulnerable to falsification. Einstein's theory of relativity, for example, received a definitive test (and confirmation) by measurements of the bending of starlight by the sun during a solar eclipse. If a theory can't be tested or falsified, it is not a scientific theory. The theory that God caused the Big Bang, for example, isn't a scientific theory, because (while it may be true) there are no observations we can make to disprove it. When a theory has withstood many tests and made many correct predictions, it becomes a scientific fact, which we can understand as a theory having such strong support that all rational people would accept it. The theories of atoms and of chemical bonds, for example, have graduated from theory to fact. Both could conceivably be shown to be wrong—all the data supporting the existence of atoms might have been deceptive—but it's highly unlikely.

So, how do Darwinism and ID compare when judged against these criteria? Let's start by looking at Darwinism.

The modern theory of evolution, called neo-Darwinism in light of 150 years of post-Darwin research, has four parts. Put simply, these are as follows:

First, evolution occurs; that is, living species are descendants of other species that lived in the past.

Second, evolutionary change occurs through the gradual genetic transformation of populations of individuals over thousands or millions of years.

Third, new forms of life arise from the splitting of a single lineage into two, a process known as speciation. This continual splitting leads to a nested genealogy of species—a "tree of life" whose root was the first species to arise and whose twigs are the millions of species living today. Trace back any pair of twigs from modern species through the branches and you will find that they share a common ancestor, represented by the node at which the branches meet.

And *fourth,* much of evolution occurs through natural selection. Individuals carrying genes better suited to the current environment leave more offspring, causing genetic change in populations over time which improves the "fit" of the organism to the environment. It is this improving fit that gives organisms the appearance of having been well designed for their lifestyles.

These claims don't necessarily stand or fall together. Nevertheless, evidence supporting all four began to accumulate starting with Darwin's 1859 *On the Origin of Species* and continues to inundate us today. Every bit of information we have gathered about nature is consonant with the theory of evolution, and there is not one whit of evidence contradicting it. Neo-Darwinism, like the theory of chemical bonds, has graduated from theory to fact.

What is this evidence? It is immense, so I will just touch upon what Darwin himself presented, though he had only a fraction of the evidence available today. It came from the fossil record, from curious remnants persisting in the anatomy and development of living species, and from biogeography—the geographical distribution of Earth's flora and fauna.

Let's start with the obvious place to look, the fossil record. Even in Darwin's time, there was evidence here supporting evolution, in the sequence of organisms laid down in the rocks. The deepest and oldest sediments show marine invertebrates; fish appear much later, and amphibians, reptiles, and mammals later still. Why should divine creation follow such a path, from the simple to the complex? Yet it is what we would expect with evolution. Darwin also observed that the species inhabiting any region—the living marsupials of Australia, for instance—closely resemble fossils found in the same place. This suggests that the former descended from the latter. We can trace evolutionary changes in lineages through the record: diatoms grow larger, clamshells get ribbier, horses become larger and toothier, and the human lineage evolves bigger brains, smaller teeth, and greater proficiency at walking on two legs. There are transitional forms, too—but more on those later.

Leaving behind the dead, we also find ample evidence of evolution among the living—relics that the evolutionary biologist Stephen Jay Gould dubbed "the senseless signs of history." They are many: The tooth buds developed in the embryonic stage by birds and anteaters—buds that are later aborted and never erupt—are remnants of their toothed ancestors. The tiny vestigial wings hidden under the feathers of the flightless kiwi attest to its ancestors' ability to fly. Some cave-dwelling animals have rudimentary eyes that cannot see, degenerate remnants of their ancestors' sighted ones. What creator, or guiding intelligence, would give animals such useless tooth buds, wings, or eyes?

Our bodies, too, are a palimpsest of our ancestry. The appendix is a familiar example. Less well known is the bad design of the recurrent laryngeal nerve—a nerve that runs from the brain to the larynx, helping us speak and swallow. In mammals, this nerve doesn't take a direct route but descends into the chest, loops around the aorta near the heart, and then runs back up to the larynx. It is several times longer than it needs to be; in the giraffe the nerve has to traverse the neck twice and so is fifteen feet long—fourteen feet longer than necessary! The added length makes the nerve more susceptible to injury, and its tortuous path makes sense only in light of evolution. We inherited our developmental pathway from that of ancestral fish, in which the precursor of the recurrent laryngeal nerve attached to the sixth of the gill arches—embryonic bars of muscle, nerves, and blood vessels that developed into gills. During the evolution of land animals, some of the ancestral vessels disappeared, while others were rearranged into a new circulatory system. The blood vessel in the sixth gill arch moved backward into the chest, becoming the aorta. As it did so, the nerve that looped around it was constrained to move backward in tandem. Natural selection could not create the most efficient configuration because that would have required breaking the nerve and leaving the larynx without innervation.

Look deeper and you find evidence for evolution buried in our DNA. Our genome is a veritable farrago of nonfunctional DNA, including many inactive "pseudogenes" that were functional in our ancestors. Why do humans, unlike most mammals, require vitamin C in their diet? Because primates cannot synthesize this essential nutrient from simpler chemicals. Yet we still carry all the genes for synthesizing vitamin C. The gene used for the last step in this pathway was inactivated by mutations 40 million years ago, probably because it was unnecessary in fruit-eating primates.

Is this really the best an "intelligent" designer can do? IDers claim that arguments for evolution based on inefficiencies or vestigial structures are unscientific because they supposedly include an unjustified theological assumption that a designer would not create such structures. But IDers are missing the point here. The evolutionary argument is that these imperfections and inefficiencies *make sense only if one assumes that evolution has occurred!* They comport with creationism only if you believe that the creator deceptively designed all organisms to delude us into thinking that they had evolved.

And finally, what of biogeography? This yields some of the most powerful evidence for evolution. It was Darwin's genius in the *Origin* to show that the distribution of plants and animals made sense only by assuming that species had evolved and split into additional species. Let's take his argument about the wildlife of oceanic islands—islands that, like the Galápagos and Hawaii, were never connected to continents but arose, bereft of terrestrial life, from beneath the sea. Compared with continents or continental islands, these islands have bizarrely unbalanced flora and fauna—unbalanced in that they are missing or impoverished in many types, while others (especially plants, insects, and birds) are present in profusion, consisting of clusters of numerous similar species ("radiations"). Hawaii, for example, has no native terrestrial mammals, reptiles, or amphibians but has large radiations of fruit flies and silversword plants. One third of the world's 2,000 species of fruit flies are found on the archipelago, although it makes up only 2 percent of the land on Earth.

Moreover, the animals and plants inhabiting an oceanic island are most similar to those species found on the nearest mainland, often despite great differences in habitat. Darwin's clinching point was this: The kinds of wildlife commonly found on oceanic islands are those that could get there easily, carried by winds, ocean currents, or other animals. Clearly, novel species on oceanic islands descend from those that were able to colonize from the nearest mainland and subsequently evolved and speciated on the islands. Only unplanned evolution makes sense of all these observations of island biogeography. No theory of design or creation even begins to explain them.

Darwin had strong evidence for evolution but no direct evidence for natural selection as the process by which it occurs. He relied on logical argument—the existence of variation in populations and its influence on reproductive success, from which natural selection followed inevitably—and on analogy with the artificial selection used by breeders to produce forms as diverse as Chihuahuas and St. Bernards from wolves, and cauliflower and Brussels sprouts from wild cabbage, within a mere 1,000 years or so.

But vast amounts of evidence have accumulated since Darwin's time. Biologists have now observed hundreds of cases of natural selection, beginning with the well-known examples of bacterial resistance to antibiotics, insect resistance to DDT, and HIV resistance to antiviral drugs. Natural selection accounts for the defense of fish and mice against predators via camouflage and for the adaptation of plants to toxic minerals in the soil. And the strength of selection observed in the wild, when extrapolated over long periods, is more than adequate to explain the diversification of life on Earth.

Neo-Darwinian evolution passes with flying colors the test of a scientific theory as an explanatory framework for wide-ranging evidence. What a remarkably elegant theory it is, and what a vast body of evidence it explains! It makes sense of data from fields as diverse as paleontology, biogeography, embryology, anatomy, and molecular biology. The geneticist Theodosius Dobzhansky famously declared, "Nothing in biology makes sense except in the light of evolution." One might add that nothing in biology makes sense in the light of intelligent design.

But can neo-Darwinism make predictions? And is it falsifiable? Yes, and yes again. As a historical science, it cannot be expected to foretell the future in detail. But it can, nevertheless, make powerful predictions.

Darwin himself made predictions from his theory. The age of the earth was unknown in his time, but he predicted that it would be old, to allow time for evolution to produce the existing diversity of life. We now know that Earth is 4.6 billion years old. He also predicted that if plants on oceanic islands were descended from those on continents, the seeds of continental plants should be able to survive prolonged immersion in seawater, and he confirmed this prediction with experiments described in the *Origin*.

Developments in biology after Darwin have served to confirm other predictions of evolutionary theory: For example, in order for natural selection to operate, there must be plenty of heritable variation in wild populations of plants and animals. We now know the source of this variation (a problem that baffled Darwin): mutations in DNA. Research in the past century has uncovered ample genetic variation for nearly every trait in every species studied.

We also now understand that natural selection involves the differential reproduction of genes. That means there should be examples of selection that benefit the genes themselves and not their carriers. Recent studies have thoroughly confirmed this prediction. In the production of eggs and sperm, for example, there is normally a 50 percent probability of each of our two gene copies going into each gamete. But there are some "selfish genes" that kill their partners and so get overrepresented in eggs and sperm. This observation does not follow obviously from the view that organisms were intelligently designed.

Neo-Darwinism predicts genetic lineages from ancestors to descendants. Darwin was concerned by the absence of transitional fossils in his day but attributed such "missing links" to the imperfection of the fossil record. He predicted that with time such forms would be found. Sure enough, we now have transitional forms linking major groups, including fish with tetrapods,[2] dinosaurs with birds, reptiles with mammals, and land mammals with whales. This is strong confirmation of evolutionary theory and destroys the creationist notion of unchanging, divinely created species (a milder form of which is advocated by ID).

Neo-Darwinism is thus a robust scientific theory, explaining a vast body of evidence, generating predictions that have been amply confirmed, and vulnerable to falsification, but showing itself more than capable of withstanding all scrutiny so far. Most of modern biology rests on its foundations.

And now to ID: How does it fare when measured by the same criteria? But let's pause for a moment. IDers complain that critics misunderstand and caricature their theory's claims, so we must first be sure we are representing

---

[2] See Neil Shubin's essay in this volume for further information on this very point.

them fairly. In the past, I have been criticized by IDers for asserting that their theory is both untestable and false. "How can that be?" they ask. The answer lies in the aforementioned ambiguity with which they present it. Torn between conflicting needs to satisfy the requirements of the First Amendment and the beliefs of their evangelical Christian supporters, IDers offer diverse (and often mutually incompatible) versions of their theory.

The theory of intelligent design oscillates between two poles that I will call the "weak" and "strong" forms. The weak form is how advocates of ID present their views in court, and the strong form is what most IDers really believe and admit to religious audiences—although its tenets are apparent in *Of Pandas and People*, the official ID textbook designed to supplement high school biology courses. The weak form of ID is untestable and tautological, while the strong form is testable in part and false. Both forms criticize neo-Darwinism for supposed flaws, but these criticisms have been shown to be misguided.[3] Let's also be clear about what ID is not. It is not *theistic evolution*—the view that God got the ball rolling with the Big Bang (or even earlier) and has kept his hands off the universe ever since. In this view, evolution occurred as part of God's plan but also according to the laws of physics and biology and without supernatural intervention. Theistic evolution is simply neo-Darwinism with the addendum that "God started it." IDers

---

[3] See, for example, Robert T. Pennock, *Tower of Babel: The Evidence Against the New Creationism* (Cambridge, MA: MIT Press, 1999), Kenneth R. Miller, *Finding Darwin's God: A Scientist's Search for Common Ground Between God and Evolution* (New York: HarperCollins, 1999), and H. Allen Orr, "Devolution: Why intelligent design isn't," *The New Yorker*, May 30, 2005.

reject this notion, because, they say, it comes with all their perceived weaknesses of neo-Darwinism.[4]

The weak form of ID consists of four assertions: First, some features of organisms don't just *appear* designed but *were* designed by an intelligent agent. Second, these features are "irreducibly complex"—that is, they could not have evolved in a stepwise fashion with each step conferring an adaptive advantage, so they cannot be the result of natural selection. Third, other features did evolve by natural selection; organisms are mixtures of some traits that were produced by intelligent design and others that evolved by natural selection. And fourth, nothing is known or can be known about the nature of the designer or the designer's goals and methods. As Lehigh biochemist Michael Behe, one of ID's principal spokesmen, put it in *Darwin's Black Box* (1996):

> Features that strike us as odd in a design might have been placed there by the designer for a reason—for artistic reasons, to show off, for some as-yet-undetectable practical purpose, or for some unguessable reason—or they might not . . . The designer might have multiple motives, with engineering excellence oftentimes relegated to a secondary role . . . Yet the reasons a designer would or would not do anything are virtually impossible to know unless the designer tells you specifically what those reasons are.

And this from *Of Pandas and People*:

[4] See W. A. Dembski, "What Every Theologian Should Know About Creation, Evolution, and Design." *Center for Interdisciplinary Studies Transactions*, 1995, 3(2):1–8.

> An intelligent designer might reasonably be expected to use a variety . . . of design approaches to produce a single engineering solution, also. Even if it is assumed that an intelligent designer did indeed have a good reason for every decision that was made, and for including every trait in each organism, it does not follow that such reasons will be obvious to us.

Does this weak form of ID qualify as a scientific theory? Consider first the evidence. If any truly "irreducibly complex" adaptations exist, then any adaptations that could not have evolved just by natural selection are obvious candidates for design. Using this strategy, IDers can then point to *any* adaptation whose evolution we do not yet understand and deem this to be "proof" of design. (This is what Richard Dawkins, elsewhere in these pages, calls "the argument from personal incredulity.") But many structures that creationists once thought irreducibly complex—the vertebrate eye; the mammalian jaw; even that superannuated poster child of ID, the bacterial flagellum—now have a scientific explanation. The argument from irreducible complexity boils down to the obvious fact that such features are always unexplained until they have been explained. Weak ID is unable to point to any features of organisms that cannot, at least in principle, be explained by evolution. And what of testability and falsifiability? Weak ID asserts that organisms contain a mixture of traits, some designed and some evolved. The designer is not limited to irreducibly complex traits; he, she, or they could have created *any* trait, even those that look as if they had evolved. Add to that the further declaration that the designer's motives and methods (and, indeed, even attitude and mental capacities) are unknown, and anything goes! An example of poor design? Maybe the intelligent designer had a bad day.

In sum, weak ID is untestable, unfalsifiable, and devoid of scientific content.

Perhaps the strong form will fare better. Strong ID shares the first three assertions of weak ID but reverses the fourth and makes some additional claims. According to strong ID, the identity of the designer is not a mystery. Forget the possibility of the designer being a race of hyper-intelligent extraterrestrials. He is very definitely identified as the Christian God. Moreover we do know something about the designer's characteristics and methods. He is intelligent, naturally, and being the Christian God, also wise and benevolent, generally bestowing on organisms the best possible features. As the philosopher William A. Dembski writes in his 1999 book, *Intelligent Design: The Bridge Between Science and Theology*, "The Hebrew notion of *hesed*, God's lovingkindness [sic], steadfastness, and tender mercy, all find their guarantee in God's role as a creator."

As for those cases of poor design, they can be attributed to the biblical fall of man. It's hard to believe that IDers think features like our appendix can be traced back to the misdemeanors of Adam and Eve, but it's true. In a later book, *The Design Revolution* (2004), Dembski notes:

> One of my critics asks, "What might be the intelligent purpose for creating species doomed to extinction? Or why would an intelligent designer create humans with spines poorly adapted for bipedal locomotion?" If we think of evolution as progressive in the sense that the capacities of organisms get honed and false starts get weeded out by natural selection over time, then it seems implausible that a

> wise and benevolent designer might want to guide such a process. But if we think of evolution as regressive, as reflecting a distorted moral structure that takes human rebellion against the designer as a starting point, then it's possible a flawless designer might use a very imperfect evolutionary process as a means of bringing a prodigal universe back to its senses.

And this from Dembski's address to the Fellowship Baptist Church of Waco, Texas:

> [A]nother thing I think we need to be aware of is that not every instance of design we see in nature needs to be directly attributed to God. Certainly as Christians we believe there is an angelic hierarchy—it's not just that there's this physical material world and there's God. There can be various hierarchies of intelligent beings operating. God can work through what can be called derived intelligences—processes which carry out the Divine will but maybe not perfectly, because of the fall.[5]

As for the designer's methods—why, miracles, of course! In strong ID, the mechanism for producing designed features is explicitly supernatural. As Dembski writes in *Intelligent Design*, "Algorithms and natural laws are in principle incapable of explaining the origin of CSI [complex specified information] . . . Indeed algorithms and natural laws are ideally suited for transmitting already existing CSI . . . [W]hat they cannot do is explain its origin." But how

---

[5] Dembski's "Lesson on Intelligent Design #6" may be heard online at: http://www.fellowshipbcwaco.org/SundaySchool/Dembski_03_07_04/Bill_Dembski_03_07_04.mp3).

are these miracles supposed to take place? Without an answer, ID lacks any mechanistic basis comparable to neo-Darwinism's natural selection. In any case, neither this tenet nor the previous three are scientific claims; they are theological suppositions that cannot be tested.

But strong ID does make four further claims that are more amenable to scientific testing. Unfortunately for the IDers, they are false, and have been known to be so for a long time. Let's deal with them one at a time.

(1) *Major groups of organisms originated suddenly, without ancestors.*

This "abrupt appearance theory" is discussed in ID pioneer Phillip Johnson's *Darwin on Trial* (1993), and in *Of Pandas and People. Pandas* claims that the fossil record shows that "fully formed organisms appear all at once, separated by distinct gaps." This claim has long been a staple of creationism, being concordant with Genesis I and II.

IDers cite two lines of evidence for their abrupt-appearance theory. One is the so-called Cambrian explosion, the relatively rapid appearance of many multicellular life-forms about 540 million years ago. But the "explosion" lasted at least 10 million to 30 million years, and the animals that appear in the fossil record were preceded by simpler ones and followed much later by the gradual appearance of other major types.[6] There is simply no evidence that all major groups of organisms originated in one instant.

The second line of supposed evidence is the arrival in the fossil record of some groups, such as rabbits and

---

[6] See, for example, S. Conway Morris, *The Crucible of Creation: The Burgess Shale and the Rise of Animals* (New York: Oxford University Press, 1998) and Miller, *Finding Darwin's God.*

bats, as distinct new types, without obvious fossil ancestors. More generally, ID has also attacked the Darwinian claim of common ancestry, asserting that we lack fossils showing relatedness between major groups. *Pandas*, for example, claims that "we cannot form a smooth, unambiguous transitional series linking, let's say, the first small horse to today's horse, fishes to amphibians, or reptiles to mammals." But as we have already noted, these transitions (and others) are now well documented with fossils and appear exactly at the point in the fossil record predicted by evolution.

There are indeed some groups of animals, including rabbits and bats, that appear suddenly in the fossil record, but given the incompleteness of that record this is hardly evidence for creation. After all, in Darwin's time we had no transitional fossils at all. As the paleontologist George Gaylord Simpson remarked in *The Major Features of Evolution* (1953), using the absence of transitional fossils as evidence for creationism is equivalent to claiming that such fossils are always missing until they have been found.

(2) *Evolution by natural selection does occur, but only within "kinds" of organisms. There are no evolutionary transitions between "kinds."*

IDers admit to microevolution, involving natural selection within a "kind," and even to speciation within a kind. But they claim that macroevolution—that is, evolution *between* kinds—comes about only through supernatural intervention. As I have shown elsewhere, this assertion derives directly from the Genesis story describing God's creation of different "kinds" of plants and animals.[7] But

---

[7] Jerry A. Coyne, "The Faith That Dare Not Speak Its Name: The Case Against Intelligent Design." *The New Republic*, Aug. 22, 2005.

what, in scientific terms, are the "kinds"? Are they equivalent to the biological phyla? Orders? Families? Something else? Nobody knows. But one thing that all creationists (including their ID brethren) do claim to know is that humans occupy their own "kind" and thus could not have evolved from other primates.[8] There is no theoretical reason for limitations to evolution within unspecified "kinds," and we have plenty of evidence of the common ancestry of major groups (including human beings)—not only from transitional fossils but also from embryology, comparison of genomes, and the existence of vestigial organs.

(3) *We do not know if the earth is young or old.*

Both young-earth creationists (who believe the earth to be just 6,000 years old) and old-Earth creationists (who accept modern geological dating) march under the banner of ID. Rather than alienate either group, ID proponents take no official position on this point. *Pandas* merely notes that design proponents "are divided on the issue of the earth's age. Some take the view that the earth's history can be compressed into a framework of thousands of years, while others adhere to the standard old earth chronology." Well, what is the truth? The scientific evidence supporting an ancient Earth is substantial and comes from several independent methods of dating. It is not doubted by any serious scientist. If IDers cannot agree about something as fundamental as the age of the earth, how can they be trusted to present any reliable or coherent view about nature?

---

[8] For an argument for the distinctness of the human "kind," see W. A. Dembski's "Reflections on Human Origins," at http://www.designinference.com/documents/2004.06.Human_Origins.pdf.

(4) *Natural selection is not powerful enough to build complex organisms.*

This one is as old as the *Origin*. Dembski, however, has updated it with fancy mathematical trimmings in writings such as his book *No Free Lunch: Why Specified Complexity Cannot Be Purchased Without Intelligence* (2001). But his assumptions and mathematical arguments have all been decisively refuted in a number of scholarly critiques.[9] We simply have no convincing arguments or proofs showing that over the 3.5 billion years of evolution, natural selection could not yield complex organisms or traits. What's more, theoretical work shows that natural selection can evolve complex organs such as the eye in a remarkably short time.[10]

Does anybody, then, really want this theory taught to their children as science? The first four tenets of strong ID are religious claims, while the other four have long been falsified. In the end, the theory of intelligent design, when it has any content at all, proves to be nothing more than a mishmash of Christian dogma and discredited science. And it is a moving target, as IDers constantly waffle to satisfy their two disparate constituencies. When pressed on their

---

[9] See for example B. Fitelson, C. Stephens, and E. Sober, "How not to detect design," *Philosophy of Science* (1999), 66:472–88; P. Godfrey-Smith, "Information and the Argument from Design," in Robert T. Pennock, ed., *Intelligent Design Creationism and Its Critics* (Cambridge, MA: MIT Press, 2001); and H. J. van Till, "*E. coli* at the No Free Lunchroom," at www.aaas.org/spp/dser/evolution/perspectives/vantillecoli.pdf.

[10] See D.-E. Nilsson and S. Pelger, "A pessimistic estimate of the time required for an eye to evolve," *Proc. Roy. Soc. Lond. B* 256 (1994):53–8; and Richard Dawkins, "The eye in a twinkling," *Nature* 368 (1994):690–91.

claims, they always revert to the weak theory. But make no mistake: What they want (as evidenced by the Wedge Document) is for the strong form, with its religious baggage, to ultimately replace the teaching of evolution in schools.

To sum up, how do evolution and ID compare as competing explanations of nature? Here are the things that evolution can explain but intelligent design cannot: the fossil record, biogeography, the existence of vestigial organs and pseudogenes, selection for genes *within* an organism, elements of poor or suboptimal design, and many other features of development. Here is what intelligent design can in principle explain and evolution cannot: any adaptation that could not possibly have evolved by natural selection. No such adaptations are known.

Evolution operates by well-known and proved mechanisms, including natural selection. Intelligent design invokes supernatural processes whose details are never spelled out and thus not subject to scientific investigation. Since 1973, more than 100,000 peer-reviewed papers on neo-Darwinian evolution have been published. ID is represented by just a single peer-reviewed paper, and this is a generous estimate, since that paper has been refuted.[11] The advocates of ID are well aware of their failure to do scientific research (which, after all, is the only way to gain scientific acceptance). They are embarrassed by this and constantly assure us that their research is right around the

[11] The ID paper is M. J. Behe and D. W. Snoke, "Simulating evolution by gene duplication of protein features that require multiple amino acid residues." *Protein Science* 13 (2004):2651–64; its refutation is M. D. Lynch, "Simple evolutionary pathways to complex proteins." *Protein Science* 14 (2005):2217–25.

corner. Five years ago, William Dembski proclaimed, "Give us more time, and we'll deliver on the program. That's our promise."[12]

I'm not holding my breath.

JERRY A. COYNE, an evolutionary biologist, is a professor in the Department of Ecology and Evolution at the University of Chicago. He is the author (with H. Allen Orr) of *Speciation*.

[12] W. A. Dembski, *Is Intelligent Design a Form of Natural Theology?* May 11, 2001, http://www.designinference.com/documents/2001.03.ID_ as_nat_theol.htm

LEONARD SUSSKIND

The Good Fight

SCIENCE AND RELIGION have never loved each other for very long. For centuries their relationship has been a contentious one. Like a neurotic love affair, it has been at times sweetly amorous and at times has boiled over into extreme forms of abuse. The bad times are well known—the burning of Giordano Bruno, the inquisition of Galileo, Darwin's fear of being made an outcast—but the story is not all one-sided. Like art and music, science has sometimes been the beneficiary of the Church's largesse. Galileo himself was a favorite of Pope Urban VIII until political considerations outweighed the Pope's affection. Right-wing intolerance for evolution stands side by side with Pope John Paul's address to the Pontifical Academy of Sciences on October 22, 1996, in which he expressed admiration for the theory of evolution as well as the view that evolution is compatible with Christian faith.

The contentiousness is of course not surprising. Whether or not evolution is compatible with faith, science, and religion represent two extremely different worldviews, which, if they coexist at all, do so most uncomfortably. Today, in the United States, science and religion are in an angrier struggle than at any time within living memory. In itself, an intellectual battle of ideas is not at all a bad thing. But what I and many other people find deeply disturbing is the mechanisms that drive the conflict. It seems that both sides are pawns in a bigger game, a game of politics and power.

Ugly as it is, the "culture battle" has its laughable side. Scientists and theologians debate exactly at what point life begins. One wonders if right-to-life zealots will be pushing baby buggies containing zygotes (or perhaps stem cells) dressed in miniature baby costumes. Meanwhile the literalists claim to believe that God buried dinosaur bones 6,000 years ago just to fool us. Equally absurd (well maybe not quite equally), we have been treated to the silly spectacle of the Kansas school board (and their attorneys) debating the scientific merit of one of the greatest products of the human intellect: Darwin's theory of natural selection. On the other hand, I have also heard absurdly inflated claims by scientists about a "Theory of Everything."

What is the reason for the recent upsurge of antiscientific passion? My own view is that it is, in part, a result of the anger, fear, frustration, and humiliation suffered over the years by the losers in the culture wars: those who would have kept women in the kitchen, blacks in the back of the bus, and gays in the closet. It is also a consequence of the deep and terrible universal fear of old age and death. But I don't believe these emotions, by themselves, could have created the antiscientific backlash of recent years. The fault may well lie in the ease with which these emotions can be cynically manipulated. It is pretty clear that the battle was engineered by provocateurs who may not even have wanted to win the battles they provoked. What seems much more likely, in view of the gingerly way that politicians have skirted such issues as *Roe v. Wade*, is that the provocateurs want to lose the battles and in that way keep the anger and humiliation at fever pitch.

How should scientists respond to this strategy? I have to admit that I'm conflicted about this question. One response that might be effective is to simply ignore the battle. The usual derisive treatment of the "Know Nothings" tends to whip up the fury and thus plays into the hands of the cynical political forces who know so well how to use it. Both sides, it seems, are being manipulated. So, then, what if we scientists refuse to play the game? After all, what great harm would come from teaching intelligent design in Kansas? Most likely, within a couple of years, parents worried about their children's ability to get into good universities would be petitioning their school boards for better biology classes.

Unfortunately, I suspect there is more at stake than biology textbooks in Kansas. As a longtime observer of the science-government-politics triangle, it looks to me as if there is another hidden agenda: to discredit the legitimate scientific community. A well-respected scientific community can be a major inconvenience if one is trying to ignore global warming, or build unworkable missile-defense systems, or construct multibillion-dollar lasers in the unlikely hope of initiating practicable nuclear fusion.

The stirring up of anti-intellectual sentiments is, of course, hardly a new political trick. Eisenhower-era political operatives made a laughingstock of thinking people by calling them eggheads, while the attacks of Senator Joseph R. McCarthy resulted in a deep suspicion of the motives of those scientists who, like Einstein, Bohr, and Oppenheimer, lobbied for nuclear sanity. The resulting political climate left the way open for lesser figures like Edward Teller to dominate the debate over the development of the H-bomb.

Today we have the ridiculous comedy of a Yale- and Harvard-educated U.S. president who plays to his antiscience audience by (deliberately?) mispronouncing the word "nuclear."

But the country pays a steep price for discrediting and dismissing the advice of eminent well-qualified experts. An example of great consequence is the Iraq disaster that followed when both Republicans and Democrats ignored the careful analyses of weapons inspectors Hans Blix and Mohamad El Baradei, and then compounded it by ignoring such military experts as army chief of staff General Eric K. Shinseki. Even worse disasters will result from ignoring the warnings of climatologists about global warming, or the advice of numerous scientists about how best to stem nuclear proliferation. So allowing ourselves to be ignored, discredited, and dismissed is not an option.

My own feeling is that in order to develop the right response it is important to understand the origin of the conflict. Ironically, the root cause of the hostility between religion and science probably involves primitive biological mechanisms. Consider such questions as "Why do humans and other animals instinctively and habitually divide the world into Us and Them?" And "What evolutionary benefits might there be in following irrational leaders?" And above all, "What are the Darwinian pressures that drive us all toward irrational belief systems?" These are scientific questions that can be answered if we have the courage to address them. Perhaps they hold some of the keys to understanding how fear and anger are so easily manipulated.

Some answers to these questions are more or less clear. The fear of death was hardwired into our genetic makeup in the deep past: Even worms flee danger. Without the death fear, we wouldn't survive to beget genetically similar offspring. Fear of death may be our deepest instinct—as deep or deeper than the urge to reproduce. But there are times when fear can cripple us, especially during a fight. A belief that one's god is all-powerful, or that Heaven is a place of unearthly delights, had obvious Darwinian value for our remote ancestors. (I don't mean to imply that there is a "religion gene"—only that suspending the death fear on occasion might have been a powerful aid in the spread of genes.) We have inherited these useful tools: No one today can deny that faith—the belief in the irrational—comforts billions of people aware of their own mortality.

Similarly, the ability to divide the world into "Us" and "Them" and then dehumanize "Them" would have been useful to roaming bands of hunter-gatherers. The same tendencies have been repeatedly reported in chimpanzees. They are all too often apparent in modern conflicts.

This is probably not the whole story. Evolution was not always consistent. No doubt we have other circuitry that is wired to resist, mistrust, and even fear, the irrational. After all, an irrational decision can get you killed. Humans are a complicated mix of competing instincts, which surface to varying degrees in different people and in different circumstances.

That brings me to the fundamental dilemma we face. Science—the quest for rational answers—does threaten the comforting beliefs that many people rely on. Conversely, the more scientific-minded find the whole notion of faith-based decision making threatening. Not surprisingly, we square off and see each other in the starkest of "Us versus Them" terms. This, I suspect, is the main reason it is so easy to manipulate the sentiments of both sides of the debate. We are primed for it by our own biological heritage. There is no doubt that certain forces have learned, either consciously or unconsciously, how to tap into these instincts and press our biologically based hot buttons.

How then do we—especially scientists—deal with the current mess without aggravating it? I believe we have no choice but to defend and protect the integrity and objectivity of science. That means building a broad consensus for the separation of religion and science (and religion and science education), and I think it can be attained; despite the fact that 90 percent of Americans believe in God, we still have managed to maintain such a broad consensus for the separation of church and state. But to achieve this goal—the separation of religion and science—we have to regain the goodwill of the public. Pressing the biological hot buttons of that 90 percent is a no-win game, a game we need to avoid. But it is equally important not to allow our own biological buttons to be pushed. Our overreaction is precisely the goal of the provocateurs.

About two years ago, a couple of Christian students from one of my undergraduate physics classes at Stanford asked to speak with me. It turned out that they had picked up some anti-intelligent-design vibes from one of my lectures, and they wanted to express their views. I agreed, and we met for lunch the next day. The conversation started with the two of them explaining why Darwinian mechanisms could not possibly account for the evolution of complicated systems like the eye—not without divine guidance. I knew a little about the subject, so I laid out for them a plausible sequence of evolutionary transitions from a light-sensitive patch to a genuine eye. Their reaction was to bring up another example, involving a propeller-like mechanism at the rear end of some microorganism. Unfortunately, at this point, I overreacted. Getting up on my bully pulpit, I insisted that modern science had provided complete explanations for everything and that furthermore the laws of chemistry, physics, and statistics had completely eliminated the supernatural from the evolution of life and the origin of the cosmos. My message was "Get real! Accept reality!"

The result was completely predictable: recriminations, and two students hostile to science. They had touched my buttons and I had hit theirs with a sledgehammer. The worst part was that when I thought about it, I felt I had not been 100 percent intellectually honest. I was not so sure that science really could explain everything. Certainly I could not prove it.

I recently had a chance to redeem myself, when another group of Christian students approached me. They had learned that I was writing a book that included in its subtitle the phrase "String Theory and the Illusion of Intelligent Design." Could I explain what the book was about?

Not only was my approach more effective but I am also much more satisfied that it was intellectually honest. I began by remarking that Darwinian evolution (not the topic of my book) had been very successful at explaining many things, and that biologists did not see any insurmountable obstruction in evolving complex life from simple microorganisms. I told them that there certainly were gaps in the story of evolution, but that the rules of the science game were to attempt to fill those gaps with natural explanations. The evolution of the eye came up again, and again I explained it.

But I also told them something else: that there were things that science didn't, and might never, fully understand, that quantum mechanics had led serious physicists to question the relation between mind and matter. I expressed the opinion that the mechanisms that underlie consciousness were obscure and not at all certain. How, then, could science prove that consciousness did not survive death? I told them that science could not prove any such thing.

Then we talked about the subject of my book: the origin of the universe, the laws of physics, and the fantastic fine-tuning needed for life to have evolved. I explained the current views on the anthropic principle. Scientists believe that the universe is extremely big and also very diverse, most of it hopelessly hostile to life and uninhabitable. But here and there some small pockets happen, by chance, to be more conducive to life, and that's where life forms. They quickly realized that this was not a view that supported intelligent design.

But I also quoted from the last chapter of my book:

> And what about the biggest questions of all: Who or what made the universe, and for what reason? Is there a purpose to it all? I don't pretend to know the answers. Those who would look to the Anthropic Principle as a sign of a benevolent creator have found no comfort in these pages. The laws of gravity, quantum mechanics, and a rich landscape together with the laws of large numbers are all that's needed to explain the friendliness of our patch of the universe.
>
> But on the other hand, neither does anything in this book diminish the likelihood that an intelligent agent created the universe for some purpose. The ultimate existential question, "Why is there Something rather than Nothing?" has no more or less of an answer than before anyone had ever heard of string theory. If there was a moment of creation, it is obscured from our eyes and our telescopes by the veil of explosive inflation that took place during the prehistory of the Big Bang.

One student then asked me a question that I knew was coming: "Do you believe in God?" My answer was simply this: "No, I personally don't. But I have numerous friends—eminent scientists—who do believe that an intelligence must have been involved in creation." Then I added, "However, all of us *do* science in the same way. We all take it for granted that science is the attempt to explain as much of the world as we can by natural mechanisms."

This time it would appear that no biological buttons were pushed. If not converts to atheism, these students left my office with no hostility toward science, at least for the moment.

My final advice is to forget arguing with those benighted zealots who would prefer that intellectual history had ended in the fifteenth century. There is no point in trying to convince the hard-core creationists—or, for that matter, the masters of manipulation. The real challenge is to reach out to the majority, to those sensible people who have been jerked around by conflicting ideologies and don't know what to think.

LEONARD SUSSKIND is the Felix Bloch Professor of theoretical physics at Stanford University and is a member of the National Academy of Sciences. His current research includes work on string theory, elementary particle physics, black holes, and cosmology. He is the coauthor (with James Lindesay) of *An Introduction to Black Holes, Information and the String Theory Revolution: The Holographic Universe* and the author of *The Cosmic Landscape: String Theory and the Illusion of Intelligent Design.*

DANIEL C. DENNETT

The Hoax of Intelligent Design and How It Was Perpetrated

IN THE SUMMER of 2005, the intelligent-design hoax, many years in the making, blossomed spectacularly. Over the course of a few weeks, its victims made headlines around the world:

(1) Cardinal Christoph Schönborn, Roman Catholic archbishop of Vienna, published an op-ed piece in *The New York Times* of July 7, proclaiming the Catholic Church's disavowal of the neo-Darwinian theory of evolution by natural selection.

(2) A month later, President George W. Bush announced at a press conference that he was in favor of teaching schoolchildren about intelligent design, saying that "part of education is to expose people to different schools of thought."

(3) Shortly afterward, Senator Bill Frist (R., Tenn.), the Senate majority leader, made the same point to reporters after a Rotary Club meeting in Nashville. Teaching both intelligent design and evolution "doesn't force any particular theory on anyone," Frist said. "I think in a pluralistic society that is the fairest way to go about education and training people for the future."

Where did nonscientists like Cardinal Schönborn, President Bush, and Senator Frist get the idea that a proposition favored by their religious perspectives is also a serious contender within science that ought to be taught?

These declarations were a public-relations coup for a well-organized group of conservative religious activists who are intent on persuading the American public that there is a significant controversy within biology about the status of the theory of evolution by natural selection. The challenger to the scientific establishment, they suggest, is the hypothesis of intelligent design. In response to their vigorous campaign, many Americans have come to think that ID is a legitimate school of thought in biology, worthy of study in classrooms. Editorials and opinion pieces have appeared by the dozens, many of them surmising that, in the interest of fairness and open-mindedness, perhaps a discussion of ID does belong in the schools. To some people, banning ID from the biology classroom smacks of censorship.

Biologists have been quick to respond, issuing incisive rebuttals to the various claims about the scientific integrity of ID, but these denunciations create the impression that an elitist scientific establishment is smothering an underdog, and a virtuous and plausible one at that—a theory quite literally "on the side of the angels."

Could so many good people have been taken in by a hoax? Wouldn't that be impossible? No. Here's how one of the most ingenious hoaxes in the history of science has been perpetrated.

First, imagine how easy it would be for a determined band of naysayers to shake the world's confidence in quantum physics ("How weird it is!") or Einsteinian relativity. In spite of nearly a century of instruction and popularization by physicists, few people ever really get their heads around the concepts involved. Most people eventually cobble together a justification for accepting the assurances of the experts: "Well, they pretty much agree with one another, and they claim that it is their understanding of these strange topics that allows them to harness atomic energy and make transistors and lasers, which certainly do work . . ."

Fortunately for physicists, there is no powerful motivation for such a band of mischief-makers to form. Physicists don't have to spend much time persuading people that quantum physics and Einsteinian relativity have been established beyond all reasonable doubt.

With evolution, however, it's different. The fundamental scientific idea of evolution by natural selection is not just mind-boggling; natural selection, by executing God's traditional task of designing and creating all creatures great and small, also seems to deny one of the best reasons we have for believing in God's existence. So there's plenty of motivation for resisting the assurances of the biologists. Nobody is immune to wishful thinking. Over the years, we've developed and tested the scientific methods needed to protect ourselves from our own credulity, but we've also found ingenious ways to fool ourselves and others. Some of the methods used to exploit these urges are easy to analyze; others take a little more unpacking.

A creationist pamphlet sent to me some years ago had an amusing page in it, purporting to be part of a simple questionnaire:

Test Two:

Do you know of any building that didn't have a builder? [YES] [NO]

Do you know of any painting that didn't have a painter? [YES] [NO]
Do you know of any car that didn't have a maker? [YES] [NO]

If you answered YES for any of the above, give details:

Take that, you Darwinians! The presumed embarrassment of the test-taker when faced with this task perfectly expresses the incredulity many people feel when they confront Darwin's great idea. It seems obvious, doesn't it, that there couldn't be any designs without designers, any such creations without a creator. Well, yes, until you look at what contemporary biology has demonstrated beyond all reasonable doubt: that natural selection—the process in which reproducing entities must compete for finite resources and thereby engage in a tournament of blind trial and error, from which improvements automatically emerge—has the power to generate breathtakingly ingenious designs.

Take the development of the eye, which has been one of the favorite challenges of the advocates of intelligent design. How on Earth, they ask, could that engineering marvel be produced by a series of small, unplanned steps? Only an intelligent designer could have created such a brilliant arrangement of shape-shifting lens, aperture-adjusting iris, and light-sensitive image surface of exquisite sensitivity, all housed in a sphere that can shift its aim in a hundredth of a second and send megabytes of information to the visual cortex every second for years on end.

But as we learn more and more about the history of the genes involved and how they work—all the way back to their predecessor genes in the sightless bacteria from which multicelled animals evolved more than half a billion years ago—we can begin to tell the story of how photosensitive spots gradually turned into light-sensitive craters that could detect the rough direction from which light came and then gradually acquired their lenses, improving their information-gathering capacities all the while.

We can't yet say what all the details of this process were, but real eyes representative of all the intermediate stages can be found, dotted around the animal kingdom, and we have detailed computer models to demonstrate that the creative process works just as the theory says. All it takes is a rare accident giving one lucky animal a mutation that improves its vision over that of its siblings; if this improvement helps it to have more offspring than its rivals, this gives evolution an opportunity to raise the bar and ratchet up the design of the eye by one mindless step. And since these lucky improvements accumulate—this was Darwin's insight—eyes can automatically get better and better and better, without any intelligent designer.

Brilliant as the design of the eye is, it betrays its origin with a telltale flaw: The retina is inside out. The nerve fibers that carry the signals from the eye's rods and cones (which sense light and color) lie on top of them and have to plunge through a large hole in the retina to get to the brain, creating a blind spot. No intelligent designer would put such a clumsy arrangement into a camcorder, and this is just one of hundreds of accidents frozen in evolutionary history that confirm the mindlessness of the process.

If you still find Test Two compelling, a sort of cognitive illusion that you can feel even as you discount it, you are like just about everybody else in the world; the idea that natural selection has the power to generate such sophisticated designs is deeply counterintuitive. Francis Crick, one of the discoverers of the structure of DNA, once jokingly credited his colleague Leslie Orgel with "Orgel's Second

Rule": Evolution is cleverer than you are. Even the most experienced evolutionary biologists are often startled by the power of natural selection to "discover" an "ingenious" solution to a design problem posed by nature.

When evolutionists like Crick marvel at the cleverness of the process of natural selection, they are not acknowledging intelligent design! The designs found in nature are nothing short of brilliant, but the process of design that generates them is utterly lacking in intelligence of its own. This ambiguity between process and product is built right into the word "design," and many evolutionists choose to say that what Darwin proved was that there *is* no design in nature—because he showed that there is no foresighted, intelligent designer. Perhaps this is the way we ought to use the word—to refer to the process, not the product—but perhaps not, and confusion over this has led to a lot of mischief. It permits the ID ideologues to exploit the ambiguity by inviting everybody to see the quite obvious evidence of design (in the product sense) and then encouraging them to conclude that this is evidence of a design *process* that is intelligent.

Who falls for this? Many people do. A particularly clear instance of the fallacy appears in Cardinal Schönborn's essay:

> The Catholic Church, while leaving to science many details about the history of life on earth, proclaims that by the light of reason the human intellect can readily and clearly discern purpose and design in the natural world, including the world of living things.

This is true in the product sense ("Look at all the brilliant designs in nature!") and false in the process sense. The

Cardinal's conclusion is that the presence of a finished product—a fully evolved eye, for instance—is evidence of an intelligent-design process.[1] This is of course a tempting conclusion, but it is just what evolutionary biology has shown to be mistaken. Yes, eyes are for seeing, but these and all the other functional organs in the natural world can be generated by processes that are themselves without purpose and without intelligence. This is hard to understand, but so is the idea that colored objects in the world are composed of atoms that are not themselves colored, and that heat is not made of tiny hot things. Sometimes the truth is far from obvious and hard to keep in focus.

The campaigners for intelligent design have become adept at feeding off the difficulty of this idea, by starting with a straightforward counterclaim and then retreating into the fog of technical confusion when their counterclaims are refuted. For instance, the philosopher William Dembski, one of the two most prominent ideologues of the ID movement, has attempted to argue that a particular sort of design product does require an intelligent designer, and that the designs found in nature include such products, but his various expressions of the argument to date, which depend on some rather abstruse mathematical formula-

---

[1] Although Cardinal Schönborn stated that his view was the "official stance" of the Roman Catholic Church, a few days later the director of the Vatican Observatory, Father George Coyne, SJ, contradicted the Cardinal in "God's Chance Creation," an essay published in the British Catholic periodical *The Tablet* (August 6). Coyne, an astronomer from Arizona, argued that a good Catholic can hold the belief that life "evolved through a process of random genetic mutations and natural selection." Since then, some Catholics have branded Coyne a heretic, while others have defended him. The Vatican has not yet clarified the situation.

tions, have been shown to be technically flawed.[2] Few, if any, theoreticians give his project any hope of success, since the flaws they have uncovered are central to his thesis.

Of course the experts might be mistaken; once in a blue moon, a renegade comes along and overthrows a portion of what had been thought to be established science. How can nonscientists assess their own judgment in this case? Not by trusting wishful thinking. If you think you can *just see* that Dembski must be onto something even though you can't follow the mathematics, you are falling right into the trap. (Can't you *just see* that Einstein has to be wrong about time? Your hunches here aren't to be trusted; this is where the discipline of careful scientific argument and experiment must come into play.) Perhaps, then, you should wait with bated breath on the sidelines while the experts duke it out in the scientific arena. This would be fine, except that Dembski has left the playing field and is appealing directly to the spectators, instead of contending with the scientists on their own terms.

In his trade books, magazine articles, and popular lectures, Dembski makes it appear that there is scientific controversy—but there isn't, as we can see by comparing his path with others. Genuine scientific controversies about evolution abound. In just about every subfield, there are challenges to various relatively well-established hypothe-

[2] See, for instance, B. Fitelson, C. Stephens, and E. Sober, "How not to detect design," *Philosophy of Science* (1999) 66: 472–88; Daniel Wolpert, "William Dembski's treatment of the No Free Lunch theorems is written in jello," *Mathematical Reviews*, available at http://www.talkreason.org/articles/jello.cfm; and Thomas D. Schneider, "Rebuttal to William A. Dembski's posting and to his book *No Free Lunch*," at http://www.lecb.ncifcrf.gov/~toms/paper/ev/dembski/rebuttal.html.

ses. The legitimate way to stir up such a storm is to come up with an alternative hypothesis that makes a prediction that is crisply denied by the reigning theory but turns out to be true, or explains something that has been baffling defenders of the status quo, or unifies two independent theories at the cost of some element of the currently accepted view. To date, neither Dembski nor any other proponent of intelligent design has produced anything like that: no experiments with results that challenge any mainstream biological understanding; no observations from the fossil record or genomics or biogeography or comparative anatomy that undermine standard evolutionary thinking. Instead, he and his cohorts use a ploy that works like this. First you misuse or misdescribe some scientist's work, provoking an angry rebuttal. Then, instead of dealing forthrightly with the charges leveled, you cite the rebuttal as evidence that there is a "controversy" to teach.

Note that the trick is content-free. You can use it on any topic. "Smith's work in geology supports my argument that the earth is flat," you say, misrepresenting Smith's work. When Smith responds with an angry denunciation of your misuse of her work, you pounce: "See what a controversy we have here? Professor Smith and I are locked in a titanic scientific debate. We should teach the controversy in the classrooms!" And here is the delicious part: You can often exploit the very technicality of the issues to your own advantage, counting on most of us to miss the point amid all the difficult details. In an essay on his Web site titled "Dealing with the Backlash to Intelligent Design," Dembski provides a candid description of his own methods, reveling in the fact that he provoked Thomas Schneider, a biologist at the National Cancer Institute, into a response that Dembski characterizes as "some hair-splitting that could only look ridiculous to outsider observers." Clever! What looks to scientists—and is—a knockout objection by Dr. Schneider is portrayed to almost everyone else as ridiculous hairsplitting.

The other leading proponent of intelligent design is Michael Behe, a biochemist at Lehigh University. He claims in his book *Darwin's Black Box* (1996) that the intricate microscopic mechanisms that even simple bacteria need to stay alive and earn their living exhibit "irreducible complexity": They depend for their functioning on having all their parts present at once, and hence could not have evolved gradually. Finding an example of such irreducible complexity has been the quest of skeptics about evolution ever since Darwin's day, and so far every case—the eye (what good is 5 percent of an eye?), the wing (could it have evolved in one fell swoop, and wouldn't it have been a hindrance until it got big enough to support liftoff?), and all the others that have been promoted—has been shown to be evolvable after all, by one roundabout route or another, not irreducibly complex. Remember Leslie Orgel's Second Rule and don't make the mistake of underestimating the cleverness of evolution.

Behe's candidates are different only in being so small, and although he thinks they are irreducibly complex, not only has he not shown any of them to be irreducibly complex (and when pressed, he admits as much) but biologists have already demonstrated the evolutionary paths that almost certainly account for his cases. While evolutionists still don't know all the steps (and hence in principle could someday be proved wrong), the task of arguing credibly that these phenomena are irreducibly complex has become much more demanding in the face of this recent work, and Behe, tellingly, has not attempted to recast his cases in its light.

In fact, no intelligent-design hypothesis has even been ventured as a rival explanation of any biological phenomenon. This might seem surprising to people who think that intelligent design competes directly with the hypothesis of non-intelligent design by natural selection. But saying, as intelligent-design proponents do, "You haven't explained everything yet" is not a competing hypothesis. Evolutionary biology certainly hasn't explained everything that perplexes biologists, but intelligent design hasn't yet tried to explain anything at all.

To formulate a competing hypothesis, you have to get down in the trenches and offer details that have testable implications. So far, intelligent-design proponents have conveniently sidestepped that requirement, claiming that they have no specifics in mind about who or what the intelligent designer might be. To see this shortcoming in relief, consider an imaginary hypothesis of intelligent design that *could* explain the emergence of human beings on this planet:

> About 6 million years ago, intelligent genetic engineers from another galaxy visited Earth and decided it would be a more interesting planet if there was a language-using, religion-forming species on it, so they sequestered some primates and genetically re-engineered them to give them the language instinct and enlarged frontal lobes for planning and reflection. It worked.

If some version of this hypothesis were true, it could explain how and why human beings differ from their nearest relatives, and it would disconfirm the competing evolutionary hypotheses. We'd still have the problem of how these intelligent genetic engineers came to exist on their home planet, but we can safely ignore that complication for the time being, since there is not the slightest shred of evidence in favor of their existence.

But here is something the ID community is reluctant to discuss: No other intelligent-design hypothesis has anything more going for it than the one above. In fact, that far-fetched hypothesis has the advantage of being testable in principle: We could compare the human and chimpanzee genomes and look for unmistakable signs of tampering by those genetic engineers from another galaxy. Finding some sort of user's manual neatly embedded in the apparently functionless "junk DNA" that makes up most of the human genome would be a Nobel Prize-winning coup for the intelligent-design gang, but if they are looking at all, they haven't come up with anything to report. Ironically, Dembski's "design inference" argument is supposed to set up a surefire test for finding just such telltale signs of intelligent tinkering in the causal ancestry of phenomena, but instead of trying to demonstrate his test in action, Dembski settles for the observation, in his Web essay "In Defense of Intelligent Design," that the ID perspective "encourages biologists to investigate whether systems that first appear functionless might in fact have a function"—as if any evolutionist would disagree with that.

Am I saying, then, that Dembski and Behe and the other ID ideologues are deliberate hoaxers who know full well that their "theories" are bunkum? No. I have no doubt that Dembski and Behe fervently believe that they have seen the truth. The history of science includes many cases of deeply committed researchers who have been taken in by their own wishful thinking and are as much victims as perpetrators of the illusions they seek to spread. In some of these famous cases, the researchers in question made up data or faked experiments when their careful experiments stubbornly persisted in yielding the "wrong" results. Or they suppressed data that contradicted their pet hypotheses. Dembski and Behe haven't done this; they don't have any experiments or observations in the first place. Others have abandoned the attempt to achieve their goals within the scientific enterprise and tried instead to win the argument in the court of public opinion by misrepresenting their opponents' claims or trying to discredit them, and this is the course that both Dembski and Behe have followed.

Science is an enormously prestigious and influential institution, for good reason. The discipline of submitting your claims to the judgment of peer review, where you have to respond to your critics or withdraw or revise your assertions, is the chief antidote to wishful thinking, and when Dembski and Behe forsake science in favor of public-relations campaigns, you can conclude that they have conceded that they just don't have the goods. They would love to have scientific validation of their hypotheses, but they have given up trying to get it and are now trying to undermine the prestige of science instead. And while they are no doubt sincere in their belief that intelligent design is a legitimate alternative to evolution by natural selection, they are simply being dishonest when they deliberately misrepresent the issues, again and again and again. One example from each will make their methods clear.

Dembski quoted the paleontologist Peter Ward as saying, "If ever there was evidence suggesting Divine Creation, surely the Precambrian and Cambrian transition, known from numerous localities across the face of the earth, is it."[3] The fact that Ward used this passage as a dra-

[3] This example is drawn from Jason Rosenhouse, "Why Scientists Get So Angry When Dealing with ID Proponents," *Skeptical Inquirer*, November/December, 2005, pp. 42–5. In a similar incident, Dembski misrepresented my own work, was caught, semi-apologized in

matic introduction to an account of the evolutionary biology that explains these transitions just fine was conveniently left out of Dembski's discussion, and when this was pointed out to him, he defended his use of Ward, saying that he was not quoting Ward out of context since he had mentioned in an aside that Ward was not a supporter of intelligent design. In other words, *Gotcha!* Scientists debate their views with tremendous vigor and rhetorical panache, but cute tricks like this have no place in the arena of genuine science, and they abound in Dembski's work.

Behe wrote in his book: "Lynn Margulis is highly respected for her widely accepted theory that mitochondria, the energy source of plant and animal cells, were once independent bacterial cells." That is indeed true, and he mentioned Margulis favorably so he could quote one of her famous over-the-top anti-establishment outbursts (somewhat out of context): "Neo-Darwinism," she said in a 1990 piece in *American Zoologist*, should be seen as "a minor twentieth-century religious sect within the sprawling religious persuasion of Anglo-Saxon biology." But Behe never even discussed the implications of the work that made her famous, which was precisely to show how the astonishing complexity of the eukaryotic cells that make up the bodies of all multicellular organisms could have evolved gradually. In other words, she is famous for actually *reducing* what otherwise would look for all the world to evolution skeptics like "irreducible complexity." When Behe turned to her more recent work, which directly challenges two of his pet

---

a piece on his Web site, and then went right back to making the same misrepresentation. See "Disbelieving Darwin Discussed" at Metaviews 030. 2000.03.23 (http://www.antievolution.org/people/mgrey/IDC/21990.html).

examples of irreducibly complex subsystems, bacterial flagella and eukaryotic cilia, he dismissed it, and that of Thomas Cavalier-Smith and Eors Szathmary, as mere "word-pictures." "What is fatal, however," he wrote, "is that neither side has filled in any mechanistic details for its model." This was not strictly true back in 1996, when Behe wrote this, and in the interim many of the mechanistic details have been filled in, but you'll still find Behe (and Dembski) quoting Margulis as an eminent biologist on the side of ID, and you won't find Behe backing off his claim that the bacterial flagellum is irreducibly complex. (A good recent account is Kenneth R. Miller, "The Flagellum Unspun: The Collapse of 'Irreducible Complexity,'" in *Debating Design: from Darwin to DNA*, ed. Michael Ruse and William Dembski, Cambridge University Press, 2004. Dembski's and Behe's replies in the same volume offer further examples of the sly twisting of words, but it isn't clear whom they are trying to impress: Certainly scientists reading their essays will see through them, and nonscientists probably won't read the book.)

When ID proponents urge that we "teach the controversy," it's worth pointing out that there are plenty of substantive scientific controversies in biology that are not yet in the textbooks or the classrooms. The scientific participants in these arguments vie for acceptance among the relevant expert communities in peer-reviewed journals, and the writers and editors of textbooks grapple with judgments about which findings have risen to the level of acceptance—not yet truth—to make them worth serious consideration by undergraduates and high school students.

So get in line, IDers. Get in line behind the hypothesis that life started on Mars and was blown here by a cosmic impact. Get in line behind the aquatic-ape hypothesis, the gestural-origin-of-language hypothesis and the theory that singing came before language, to mention just a few of the enticing hypotheses that are actively defended but still insufficiently supported by hard facts.

The Discovery Institute, the conservative organization that has worked so ingeniously to put intelligent design on the map, complains that its members face hostility from the established scientific journals. But establishment hostility is not the real hurdle to the intelligent-design movement. Consider why: Literally thousands of peer-reviewed scientific articles are published every year elaborating and extending the basic theory of evolution, and most of the authors of these articles never become famous, in spite of their proven expertise. Surely a few of them would happily jump ship and risk ridicule from the establishment for the chance to become world famous as the Scientist Who Refuted Darwin. If intelligent design were a scientific idea whose time had come, young scientists would be dashing around their labs, vying to win the Nobel Prizes that are certainly in store for anybody who can overturn any significant proposition of contemporary evolutionary biology. Remember cold fusion? The establishment was virulently hostile to that hypothesis, but scientists around the world rushed to their labs in the effort to explore the idea, in hopes of sharing in the glory if it turned out to be true.

Instead of spending more than $1 million a year on publishing books and articles for nonscientists and on other public-relations efforts, the Discovery Institute should finance its own peer-reviewed electronic journal. This way, the organization could live up to its self-professed image as the doughty defenders of brave iconoclasts bucking the establishment. For now, though, the theory they are promoting is exactly what the libertarian author George

Gilder, a cofounder and longtime affiliate of the Discovery Institute, has said it is: "Intelligent design itself does not have any content."

Since there is no content, there is no "controversy" to teach in biology class. But here is a good topic for a high school course on current events and politics: Is intelligent design a hoax? And if so, how was it perpetrated? There are plenty of fascinating further details behind the sketch given here.

DANIEL C. DENNETT is University Professor and Austin B. Fletcher Professor of philosophy and director of the Center for Cognitive Studies at Tufts University. He is the author of, among other books, *Consciousness Explained, Darwin's Dangerous Idea, Freedom Evolves,* and, most recently, *Breaking the Spell.*

NICHOLAS HUMPHREY

Consciousness: The Achilles Heel of Darwinism? Thank God, Not Quite

In crossing a heath, suppose I pitched my foot against my conscious self . . .

WILLIAM PALEY, in his famous statement of the argument from design in 1800, imagined that he found a watch lying on a heath and set to wondering how it came to be there: "The inference we think is inevitable, that the watch must have had a maker: that there must have existed, at some time and at some place or other, an artificer or artificers who formed it for the purpose which we find it actually to answer, who comprehended its construction and designed its use." But, so Paley argued, "every indication of contrivance, every manifestation of design, which existed in the watch, exists in the works of nature"—notably in plants and animals. From which it follows that these works of nature must likewise have had a maker.

As we now know full well, this argument does not hold up. Charles Darwin's great achievement was to show that "every indication of contrivance, every manifestation of design" can be introduced into living things by the blind process of natural selection. What's more, as later Darwinians have discovered, natural selection has the power to create the equivalent of good design not just in living things but in every other realm where variant plans or ideas compete for survival. Engineers, for example, increasingly make use of evolutionary algorithms, based on natural selection, in the development of useful products such as the wings of airplanes. Indeed, Paley's argument may soon have to be turned on its head. For there will likely come a time in the not too distant future when a complex artifact found lying on the heath will *not* have had a maker but rather have been "grown" via a genetic algorithm.

Yet though Paley's standard argument for a designer has been effectively refuted by Darwin's theory, there is a related argument that certainly has not been—at least not yet. It is an argument championed by the codiscoverer of natural selection, Alfred Russel Wallace, although it goes back to René Descartes.

Let's grant that Paley was right to point out that it requires some kind of clever process to produce a watch. Even so, it is clear enough that watches belong to the world of normal physical phenomena. In principle, once the mechanism is explained to us, we can see how and why it works: that is, how simple non-watch materials can have been put together to function, according to the laws of physics, so as to constitute this very watch. In short, there's nothing obviously impossible to the art of watchmaking.

But suppose we were to come across an object on the heath that, as far as we can see, does not belong to the world of normal physical phenomena—something whose nature (let's not say what it is yet) is such that we cannot understand it as a mechanism or see how it could have been put together to function as it does. In short, something seemingly impossible to make. Suppose, for illustration, we were to stumble on the object in this drawing. Surely the inference would be inevitable that we were confronted by evidence not merely of an intelligent designer but of a designer with august supernatural powers.

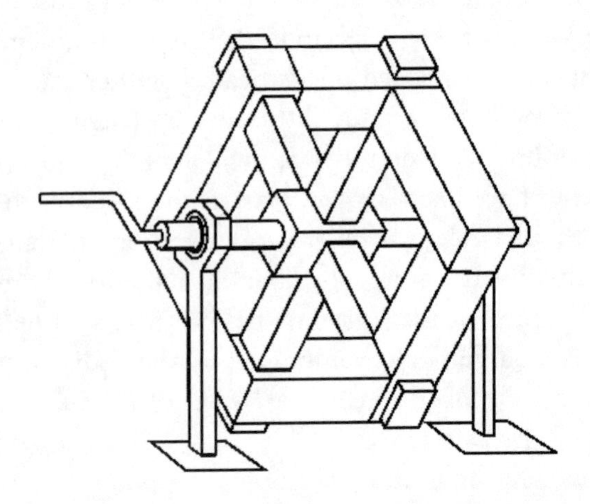

Now, on the whole, the works of nature do not present us with a challenge on this level. Living things, remarkable as they are, are nonetheless physical mechanisms, made of purely material substance. And even if there are still some puzzles about what's gone into their evolutionary design (and there really are not many such puzzles left), biology has progressed so far that we can see in almost every case how the living machinery operates. Many scientists would predict that it will soon be *every case but one*. And yet this one hard case is so central to our world, and of such personal importance, that it may be enough to call the whole scientific enterprise into question. The problem case is *human consciousness*.

Consciousness, and in particular sensory awareness, is unquestionably a phenomenon apart. As each of us has reason to observe and marvel at, at every moment of our waking lives there is *something it is like* to be us. We are the subject of feelings and sensations. We sense the heat and redness of fire, the sour tang of a lemon, the stab of pain. The space of our sensations is *our* space, the space where our selves are most at home. But this space is strangely closed off, not apprehensible by others, and beyond verbal description even to ourselves. To reduce consciousness to a mechanism would seem impossible.

The problem of consciousness has puzzled and continues to puzzle the best minds in philosophy and science, to the point of distraction. In an 1870 essay titled "The Limits of Natural Selection As Applied to Man," Wallace quotes the physicist John Tyndall:

> The passage from the physics of the brain to the corresponding facts of consciousness is unthinkable. Granted that a definite thought, and a definite molecular action in the brain occur simultaneously, we do not possess the intellectual organ, nor apparently any rudiment of the organ, which would enable us to pass by a process of reasoning from the one phenomenon to the other. They appear together, but we do not know why. Were our minds and senses so expanded, strengthened, and illuminated as to enable us to see and feel the very molecules of the brain; were we capable of following all their motions, all their groupings, all their electric discharges, if such there be; and were we intimately acquainted with the corresponding states of thought and feeling, we should be as far as ever from the solution of the problem: "How are these physical processes connected with the facts of consciousness?" The

chasm between the two classes of phenomena would still remain intellectually impassable.

The prevailing opinion at the beginning of the twenty-first century has hardly changed. The philosopher Jerry Fodor: "Nobody has the slightest idea how anything material could be conscious. Nobody even knows what it would be like to have the slightest idea about how anything material could be conscious." The psychologist Stuart Sutherland: "Consciousness is a fascinating but elusive phenomenon; it is impossible to specify what it is, what it does, or why it evolved. Nothing worth reading has been written about it." And the philosopher Colin McGinn: "Isn't it perfectly evident to you that [the brain] is just the wrong kind of thing to give birth to consciousness? You might as well assert that numbers emerge from biscuits or ethics from rhubarb."

The bottom line for many contemporary philosophers is this: If we did not already know that consciousness exists, it would be completely unnecessary to invent it. Nothing about the physical world, or about human or animal behavior, points to the possibility of there being any such thing. Consciousness cannot be deduced as a fact of life in human beings or, for that matter, in men from Mars. While scientists are well prepared to engage in SETI (the Search for Extra-Terrestial Intelligence), because they know what the manifest signs of intelligence are likely to be, they could not even begin to engage in SETC (the Search for Extra-Terrestial Consciousness), because they have nothing to go on.

Still, we do know it exists in human beings, even if we know this only from one singular, lonely case. And one case will do. To return to Paley's argument, one watch suffices to prove the necessity of a designer. So, what would Paley say to one case of consciousness: his own? Let's frame it as the story of our coming across consciousness unexpectedly out there, in our own nature. We have already noted that if we were to come across an impossible-to-create object like that in the drawing, we would have to conclude that it had been created supernaturally. But the phenomenon of consciousness gives every indication of being impossible to create. So, surely, we should not shirk the conclusion—indeed, maybe we should welcome it—that our consciousness, too, must have been created supernaturally, by a supernatural agent using superintelligent design.

Descartes set out a similar line of reasoning in his *Meditations* in 1641, although his concern was not so much with consciousness as with the "idea of God." Descartes was forever walking the heath of his own mind, and rather like Paley he discovered in his *Third Meditation* something that astonished him: that he was able to conceive of God as a perfect being. But it stands to reason, he said, that you cannot get more out of a creative process—whether it involves things or ideas—than you put into it. (This is the principle of sufficient causation: Something cannot arise from nothing.) So it should not be possible for a thinker to create the idea of perfection unless the building blocks of his thought were already perfect. However, as a mere human being, Descartes could surely not have had any perfect thoughts to start with; hence his idea of God as a perfect being was not something he could possibly have constructed by himself. Where could this impossible-to-think-up idea have come from? Descartes concluded that it had to have been miraculously implanted in him by God from above."I recognize that it would be impossible for me to exist—having within me the idea of God—were it not the case that God really existed. By 'God' I mean the very being the idea of whom is within me, that is, the possessor of all the perfections which I cannot grasp, but can somehow reach in my thought."

When stated in this abstract way—as an argument about how anyone could think an impossibly rich thought, such as the thought of God is—it may not strike you (as it has not struck most later philosophers) as terribly impressive. But if the parallel argument is made about how anyone could have evolved to experience an impossibly rich kind of experience, such as the experience of consciousness is, it becomes hard to deny its considerable force.

To make the argument as it applies to consciousness still more explicit: It seems clear, on the basis of our personal experience, that consciousness has strange, otherworldly properties. You cannot make something otherworldly out of worldly materials. So, since worldly materials are all that the human body and brain have to work with, then consciousness can't have been constructed in any normal way. Therefore, consciousness must have been specially created and planted in us.

Like it or not (I rather hope you don't), this is an argument that evolutionary scientists cannot simply run away from. There are only three reasonable ways to deal with it. One would be to claim that we are actually not conscious in the otherworldly way we think we are (as Daniel Dennett, for example, has sometimes suggested). A second would be to claim that we are actually not purely material beings (as David Chalmers has suggested). But if neither of these suits you, the third must be to acknowledge that there is something remarkable going on, God only knows what.

I think there *is* something remarkable going on. So remarkable that, as things stand, consciousness could still prove to be the Achilles heel of Darwinism, or (if I may switch metaphors) the smoking gun that gives the champions of intelligent design the evidence they seek that human beings did not evolve entirely by natural selection.

Wallace saw this, and within ten years of the publication of the *Origin of Species* he announced that the mystery of consciousness was proof (one among several) that natural selection was not enough:

> No physiologist or philosopher has yet ventured to propound an intelligible theory, of how sensation may possibly be a product of [material] organization; while many have declared the passage from matter to mind to be inconceivable. . . . *You cannot have, in the whole, what does not exist in any of the parts* [my italics (shades of the principle of sufficient causation)]. . . . The inference I would draw from this class of phenomena is, that a superior intelligence has guided the development of man in a definite direction. . . . It does not seem an improbable conclusion that . . . the whole universe, is not merely dependent on, but actually *is*, the WILL of higher intelligences or of one Supreme Intelligence.

Darwin wrote to Wallace,"I hope you have not murdered too completely your own and my child." But Darwin must surely have been worried. In one of his early notebooks he had written "The soul by the consent of all is superadded. Animals not got it."

There is clearly a case to answer here. To say the least, it suggests a potential weakness in the Darwinian account of human evolution. Yet despite the lead given by Wallace nearly 150 years ago, this is a weakness that the latter-day proponents of intelligent design have been surprisingly slow to exploit. Perversely, they have continued in the tradition of Paley, concentrating their efforts on targets such as the eye or protein synthesis, which are actually not so difficult for Darwinian theory to deal with, while leaving alone the one work of nature where the theory looks at risk. Is the explanation for their reticence that even they are shy of arguing, as Wallace and Descartes did, for super-intelligent design?

Maybe so. At any rate, before they see the potential that it holds for them, let me try to steal the case back for Darwinism by showing how consciousness could—against the odds—have evolved by natural selection as a biological adaptation. This requires several steps, all of them quite radical.

First, we have to accept the fact that consciousness, as we experience it, really is just as amazing and deeply enigmatic as it seems to be. The biologist J. B. S. Haldane once observed that the universe is not only stranger than we imagine, it is stranger than we *can* imagine. But in consciousness we find ourselves right at the center of something that, if we were not already conscious, would be far beyond what we could imagine. Let's not deny this or try to explain it away as an illusion. The paradoxical features of consciousness are precisely that with which scientific understanding has to begin.

Nonetheless, even though we should not try to explain it away, this doesn't mean we must assume that consciousness is *not* an illusion. Our starting assumption as scientists ought to be that on some level consciousness has to be an illusion. The reason is obvious: If nothing in the physical world can have the features that consciousness seems to have, then consciousness cannot exist as a thing in the physical world. So while we should concede that as conscious subjects we do have a valid experience of there being something in our minds that the rules of the physical universe don't apply to, this has to be *all it is*—the *experience of something in our minds*.

This challenge to conciousness's ontological status may dismay us, but it should not surprise us. We can, and regularly do, have the experience of being in the presence of things that don't physically exist: ghosts, for example, or mirages. More to the point, we sometimes experience things that could not possibly exist for logical reasons: There is a visual illusion called pure phi, in which we see an object moving without changing its position, and another illusion in which we see a surface as being both red and green at the same time. Such illusions are accidental errors of judgment that occur when our perceptual systems rely on rules for interpreting environmental information under conditions where those rules happen not to hold.

But is it plausible to suppose that our experience of consciousness is likewise some kind of *accidental* error? It seems too beautiful, too much of a piece—indeed, too "God-given"—to be so. Precisely so, and here's the thing. Since it is the wonderfully enigmatic features of consciousness that strike us so forcibly when we reflect on it, then perhaps it is those very features that give consciousness its role in life. And if that's the case, consciousness could have come to have these features not at all by accident but because it has been *designed to give the impression of having them*—designed by natural selection, that is. So, while our experience is indeed an error, it is one for which we have been "deliberately" set up.

If consciousness has in fact been designed to appear to have unfathomable properties, this would certainly make the task of explaining it in a conventional manner much easier. For it could have been a relatively straightforward matter for natural selection to have designed a mental phenomenon to *give the impression* of being paradoxical, otherworldly, or illogical even when there is no way to create such a phenomenon for real. You can draw a picture of the object illustrated earlier in this essay even if you cannot construct the physical object. You can devise a trick proof that 2 + 2 = 5 even if you cannot actually make two and two equal five. (Here is such a proof. "Both this and the following statement are false. Two plus two equals five." Think about it: If the first statement were true it would be self-contradictory, so it must be false; but it can only be false if the second statement is true.)

Furthermore, it might have been possible for natural selection to egg the pudding, as it were, by designing people's mental faculties so that they put an even more grandiose interpretation on their own enigmatic experience than they might otherwise have done. Thus our minds could have been specially constructed so as to make us all too ready to interpret a garden-variety paradox as a mark of the divine, or small-time consciousness as big-time consciousness of the kind that leads to belief in an immortal soul. Admittedly, this is just what Descartes and Wallace argued could not have happened: Unless you are God, you cannot design a mind to come up with ideas bigger than those already contained within it, *because you cannot have in the whole what does not exist in any of the parts*. But, with respect, we should not take these sages' word for it. Just one example can serve to contradict the principle of sufficient causation as they use it. Take the case of the idea of infinity. Could a finite human brain generate the idea of infinity? Yes, it could and does, because the idea of infinity does not have to be an infinite idea.

We can see, then, how it might have been within the

power of natural selection to design human minds to experience consciousness the way they do. But there remains, of course, a crucial proviso. The basic premise of Darwinism is that natural selection selects for traits that are biologically adaptive—leading to greater success in reproduction. So natural selection will have designed human minds to experience consciousness this way only if, in the history of our species, individuals who harbored this experience made out especially well in the struggle for survival. And how could that be? Why should our particular take on being conscious bring us biological good fortune? To answer this, the most important question, we must look objectively at what the encounter with the magically rich features of consciousness actually does for us—how, if at all, it changes human lives.

The natural history of consciousness and its effects is not a study to which scientists have yet given enough attention (although artists have been exploring it since art began). But, while there is still much to learn, I believe that if we ask the question in the right way—"How does consciousness matter? What would be missing if we didn't have it?"—a raft of answers readily becomes apparent.

To start with, without the experience of being conscious we would simply not have the sense we have that there is something substantial at the center of our psychical existence, something it's like to be us. But once we do experience it, the *Self* is there for us. A self that has this at its center is a self to be reckoned with, a self worth having. And such a self bursts with the potential to become the principle around which we organize our private mental lives.

This could be true not only for human beings but for animals as well. Perhaps whales, dogs, monkeys—maybe all mammals, all birds—have comparable experiences of consciously being there. But for human beings it evidently goes much further. For in our case we now have a self that seems to inhabit a separate universe of spiritual being. And this is something else. As the subjects of something so mysterious and strange, we humans gain new confidence and interest in our own survival—a new interest in other people, too. We begin to be interested in the future, in immortality, and in all sorts of issues to do with co-consciousness and how far consciousness extends around us.

This feeds right back to our biological fitness, in both obvious and subtle ways. It makes us more fascinating and more fascinated, more determined to pursue our lives wherever they take us. In short, more like the amazing piece of work that humans are. Lord Byron said that "the great object of life is sensation—to feel that we exist, even though in pain." That's the raw end of it. But at a more reflective level, what keeps us going, gives us courage, makes us aim high for ourselves and our children, is the feeling that as human selves we have something very special to preserve.

Matters might rest there. But in the context of writing about intelligent design, I want to leave you—and tease you—with one further idea about how consciousness can change our view of things, for which we have the evidence right before our noses.

The novelist Thomas Mann, when asked to contribute his thoughts to Edward R. Murrow's 1950s radio program *This I Believe*, came out with the following:

> In my deepest soul I hug the supposition that with God's "Let there be," which summoned the cosmos out of nothing, and with the generation of life from the inorganic, it was man who was ultimately intended, and that with him

> a great experiment is initiated, the failure of which because of man's guilt would be the failure of creation itself, amounting to its refutation. Whether that be so or not, it would be as well for man to behave as if it were so.

Mann had earlier written, in *The Magic Mountain*, "Consciousness, then, was simply a function of matter organized into life; a function that in higher manifestations turned upon its avatar and became an effort to explore and explain the phenomenon it displayed—a hopeful-hopeless project of life to achieve self-knowledge." Hopeful-hopeless it may always be. But, as we have seen with Wallace, and I would guess was true of Mann himself, and is surely true of many of the rest of us as well, it is reflection on the mysteries of consciousness—our very failure to see how consciousness "fits in"—that is the chief inspiration for the belief in special creation that Mann hugged to his soul.

Now, as Mann hints, this belief—whether true or not—may be a significant life force in its own right. For if we can indeed believe that we owe our lives to a beneficent creator who (presumably having some choice in the matter) deliberately arranged that the world he created would come to contain human beings, then this can and arguably ought to encourage us to make it our cause in life to honor the very cause of life, by living up to our creator's plan.

So, here's the irony. Belief in special creation will very likely encourage believers to lead biologically fitter lives. Thus one of the particular ways in which consciousness could have won out in evolution by natural selection could have been precisely by encouraging us to believe that we have *not* evolved by natural selection.

Anyone for "natural creationism"?

NICHOLAS HUMPHREY is School Professor at the London School of Economics and is internationally known for his work on the evolution of human intelligence and consciousness. He is the author of *The Mind Made Flesh: Essays from the Frontiers of Biology and Evolution* and *A History of the Mind: Evolution and the Birth of Consciousness*.

TIM D. WHITE

Human Evolution: The Evidence

The question of questions for mankind—the problem which underlies all others, and is more deeply interesting than any other—is the ascertainment of the place which Man occupies in nature and of his relations to the universe of things.

—T. H. Huxley,
On the Relations of Man to the Lower Animals (1861)

SUNK BETWEEN TWO imposing escarpments of broken lava rock, the triangle-shaped floor of Africa's Afar depression is a vast, barren, sparsely inhabited corner of the continent. The Awash River winds through this desolate region, fed by seasonal rains in the adjacent Ethiopian highlands. Thin ribbons of riverine forest mark the winding course of the muddy Awash. In its middle portion, the river sweeps southeast and then bends sharply north around a peninsular projection of low hills on its left bank, called Bouri by the nomadic Afar pastoralists who live there. In their language, *bouri* means dust.

In January of 1997, instruments on buoys in the Pacific off the coast of Peru began to record abnormally high seawater temperatures. The El Niño that these temperatures heralded became the strongest in more than fifty years. Its effects were global. Rains inundated the Afar, where our international research team converges every autumn to conduct investigations into the deep human past. After the deluge, the 1997 field season would yield a bumper crop of newly exposed evidence for our geologists, archaeologists, and paleontologists.

The road to Bouri's Herto village snakes along a drainage divide in the Ethiopian highlands. Rainwater draining to the east flows into the Blue Nile and ends up in the Mediterranean. West of the road, rainwater flows down the steep rift escarpment into the swamps and lakes dotting the Afar floor. The Afar is a gigantic basin, a trap where the water-transported sands and silts have accumulated to form a mile-deep layer-cake succession of ancient sediments.

When we finally managed to get the vehicles across swollen streams and deep mud, we found Herto abandoned. No herdsmen, no goats, cats, children, sheep, cows, camels, or life. Houses once covered with grass mats were now silent skeletons of curved sticks. The plain stretching away to the southeast was covered with a bright stubble of newly sprouted grass. The usual swarms of curious naked children did not materialize—it was a once-in-a-lifetime chance for a peaceful lunch at Herto. Most of the crew dispersed to look around, while the rest of us stretched a tarp from a roof rack for shelter against the harsh sun.

One investigator found a small fragment of brown and white bone. Measuring less than a centimeter thick, it fit comfortably into his palm. "Hominid," he announced. It was a piece of human skull. After lunch, I sent him back with yellow plastic pin flags to mark the location of additional fragments, and we collected evidence in the eerie silence.

The skull had been damaged postmortem. Its left side had emerged first from the sands, and the broken, fossilized pieces of its vault had been scattered by passing livestock and the whirlwinds that regularly sweep this landscape. A pattern emerged as the yellow pin flags were placed. The spread of flags narrowed and concentrated on a spot about the size of a football.

In a forensic case, investigators tiptoe through the crime scene so as not to disturb the smallest clue. Blood-spatter, fibers, fingerprints, DNA, and many other traces provide the clues needed to probe the past and solve the crime. The past probed by forensic scientists is usually not very old. In paleontological science, the past can be millions of years deep. But in both sciences, every detail is documented photographically and trace evidence is sealed in special containers. At Herto, the outline of a human cranium arose from the indurated sands of an ancient riverbed, as a fine paintbrush swept away the loose sand. The profile above the right eye socket was prominent; there was no accompanying lower jaw nor a trace of the other approximately 200 bones that would have comprised this individual's body in life. For fossils and forensics alike, establishing a timeline of events is crucial. Whose cranium was this? How had it come to rest in this sandstone layer at Herto? What had happened here? And when? The detective work to solve this case would not be easy; the clues were few, the case cold.

Here was a brittle, fossilized human cranium from a past that nobody could remember, exposed to the light of day by El Niño rains. To remove it, I excavated the sands around it and solidified them with superglue. We applied a separating layer of wet tissue paper, and then plaster, which would protect the fossil from further damage on the long trip back to the laboratory; we lifted it from its resting place that night, safe in its plaster jacket.

The investigative team returned to the scene the next morning, hoping to collect more evidence. A grid was established around the site and all surface objects were plotted. Excavation of the immediate scene produced no further remains, so we moved outward. Geologists traced the sandstone layer south to where it was covered by several feet of consolidated volcanic glass, which had formed when a nearby volcano erupted. Since the skull-bearing sands were below this layer, they would be older than the eruption. Archaeologists scoured the neighborhood for more outcrops of these sands, collecting thousands of stone tools and fossilized animal bones contemporary with the human cranium. Many bones were embedded nearby in the same sandstone layer, including those of a large buffalo-like animal with horns as long as your arm, an animal that had vanished from Africa long before domestic cattle arrived thousands of years ago.

Other stones within the sandstone also pointed to a great antiquity; they were finely shaped implements of obsidian and basalt, yet the nearest natural outcrops of these volcanic rocks are miles away. The Herto villagers of today do not make such stone tools, so this was obviously not the grave of a local inhabitant. And there was no disturbance of the sandstone layers encasing the cranium.

Within weeks, the fieldwork ended. The evidence, including the human cranium in its plaster jacket, was carefully packed and transported to the National Museum of Ethiopia, in Addis Ababa. Five years of laboratory investigation would ensue before our team was ready to present its conclusions based on the evidence recovered. Removing the hardened sandstone one grain at a time from the fragile cranium alone took months. Working with a needle under a microscope, we hardened the fossil with preservative as its fragile face emerged. It was a massive, deep face, but one with a depression beneath the eye sockets matching that of modern people. The missing brain's volume was measured at 1,450 cubic centimeters—some 100 cc more capacious than that of the average modern human. The frontal bone was vertical, as were the bones forming the sides of the cranial dome. The rear of the vault showed scars indicating the attachment of heavy muscles. This was a large individual but clearly a male of our species, *Homo sapiens*.

Meanwhile, the volcanic rock samples were being fingerprinted at the Los Alamos National Laboratory in New Mexico. Each volcanic eruption has its own unique chemical composition; ours matched a layer from southern Ethiopia. Fractions of this correlative rock were melted with a laser and the released argon gas measured in a mass spectrometer. The amount of gas trapped in the rock since it cooled would correspond to the decay rate of the radioactive element potassium—hence we could calibrate the timeline: The Herto individual died at least 155,000 years ago. Earth has been home to life for billions of years. Compared with dinosaurs, which arose approximately 225 million years ago and went extinct some 160 million years later, the Herto individual was a relative youngster, but in human generations he was old enough to be someone's great-great-great- (insert about 8,000 "greats") grandfather.

In a forensic investigation, genealogical relationships are often established through DNA analysis. Unfortunately, the Herto man's DNA had decomposed. Our identification would have to rely on comparative anatomy. We compared Herto to more than 5,000 recent human crania in institutions ranging from the Smithsonian to the University of Tokyo. Even though he came from Africa, the Herto man was like no recent African. His face was too tall, the bony ridges atop his eye sockets too prominent, his neck-muscle attachments too robust. Of all modern human populations, he most closely resembled Australian aboriginals. Analysis showed his anatomy to lie at the outer limits of the large envelope of skull form among modern humans. From a paleontological perspective, this individual's skull was evidence for evolution, linking living people with their hominid ancestors.

What had his world been like? The archaeologists had most of the evidence for that. Their studies of cutmarks left by stone tools on hippopotamus and bovine bones found in the sandstone showed that these people were butchers, with a taste for mammalian flesh. Studies of the associated artifacts showed them to be adept at manufacturing the tools needed for a life of foraging in the tropical African Pleistocene.

The Herto man died of unknown causes about 8,000 human generations ago, near a freshwater river. He lived at a time when the Northern Hemisphere was locked in an ice age and a refrigerated Europe was inhabited by cave dwellers with skull shapes very different from the Herto people. Where did the Herto people come from? And what became of their descendants?

Anthropologists have long known that every culture has a traditional story about how people were created. There are Babylonian, Hindu, Cherokee, Yoruba, Maori, Norse, Mayan, and dozens of other myths. Our Western biblical origin stories are rooted in the pastoral societies of the Middle East and have been retold for centuries. Literal interpretation of the biblical account holds that humans remained basically unchanged after they were created. This interpretation was tested by the discovery of the Herto man. Here was someone whose anatomy linked him to earlier fossils that were not human. When he appeared on the cover of *Nature* in 2003, he represented new evidence for an old debate. Who was this man and what did he represent?

Answers to these questions came quickly from people wishing to accommodate the new evidence to their particular views. The Herto man, dead for some 155,000 years, ignited a controversy that pitted leading creationist ministries against each other. One group, Answers in Genesis, an Australian-based ministry whose stated aim is "to develop a biblical worldview [and] expose the bankruptcy of evolutionary ideas," contested the radiometric dating and saw the Herto man as a descendant of Adam. The other, the Progressive Creationists—whose chief spokesman is the televangelist Dr. Hugh Ross, author of *Creation and Time*—accepted the dating but called Herto a nonhuman bipedal primate. In the ensuing set-to, these parties accused each other of special pleading, sinning, distortion, and the commission of factual errors. The Progressives, noted Answers in Genesis, represented "a greater danger to the church than attacks by atheistic evolutionists from 'without the camp.' " The debate proved sterile, but itself constitutes evidence of the degree to which human fossils can upset strongly held religious beliefs.

Atop the mile-thick Afar sediment stack, the processes of deposition and fossilization continue today. Dead cows and crocodiles settle to the bottom of swamps, where they are covered with sediment. Slowly their bones turn to stone. Below them are the remains of their ever-older ancestors, a menagerie including many animals now extinct.

These fossils would remain suspended in the sediment deep below the surface of the Afar were it not for the fact that the Arabian peninsula is slowly rotating away from the African continent. The continental separation is geologically fast but too slow for humans to observe without instruments; the rift valleys created by the movement of the Arabian plate widen by only a few centimeters a year. Since the time of the Pharoahs, the Afar rift has widened about the length of a football field. Because it is one of the most tectonically and volcanically active places on Earth, we don't need a drill or a mining operation to reach the sediment layers beneath the Herto sandstone. Over the last 6 million years, bits of the thick Afar sediment pack have been thrust back to the earth's surface. The resulting patchwork of re-exposed sediments provides investigators with a variety of windows on deep time. Annually eroded by seasonal rains, they regularly yield the voluminous bony and trace evidence through which we come to know the distant past.

In chapter 6 of his *Descent of Man* (1871), Darwin wrote that "those regions which are the most likely to afford remains connecting man with some extinct ape-like creature, have not as yet been searched by geologists." But not even Darwin could have imagined that a single geological depression in the Horn of Africa would by 2005 have yielded a record of human evolution stretching across the last 6 million years.

A stone's throw from the Herto discovery site, a prominent volcanic ash dated to 250,000 years ago underlies the Herto sandstone. Below it we have found the remains of thousands of extinct animals, including some fossilized hippopotamus bones bearing cutmarks made by ancient butchers who left hundreds of their finely made handaxes to record their presence. Our surveys have failed to recover any bones of the people who made these tools, but several miles downstream, on the opposite bank of the modern Awash River at Bodo, another package of sediments has yielded handaxes even more primitive and the cranium of a man more than twice as old as the fellow who so disturbed the dueling creationists. Dated to 500,000 years before the present, this cranium from Bodo is different from that of any living human, but it is hardly the cranium of an ape. The braincase measures 1,250 cc, smaller than Herto's but much larger than any modern ape's.

Deeper in time, we have found cruder stone handaxes at Bouri, several hundred meters east of the Herto find. Here we unearthed a more heavily browed but smaller (995 cc) skullcap of yet another hominid, this time eroding from sediments including a volcanic ash a million years old. This individual is so unmistakably not human that it bears a separate species name, *Homo erectus*. The animal fossils found in these so-called Daka strata contain animals that, unlike the Herto fauna, no longer exist. Some disappeared without descendants; others persisted but changed through time. And so it goes, as you proceed into deeper and deeper sediment layers in the Middle Awash sediment trap.

Moving east of the Daka, toward the Awash River, you cross a large time gap in sedimentation, reflecting local nondeposition. Because of this sedimentation hiatus, the Middle Awash lacks a good record of what happened here between 1.0 million and 2.5 million years ago. But farther south in the African rift, sedimentation preserves abundant evidence dating from this interval. At Bouri, there are even older sediments, dated to about 2.5 million years ago. Fossils found in these sediments are removed from us by 125,000 generations. If just one genealogical lineage with all the intervening mothers and fathers were buried in pairs in foot-deep coffins, the coffin stack would be twenty-three miles deep. And this is only halfway down the time range sampled by the deep Middle Awash sediment stack!

For deep geneaology, the terrestrial geological record is unfortunately incomplete. The bodies of most mothers and fathers decay and leave no traces. It is only under special conditions, like those in the Afar, that a small fraction of mortal remains can be preserved and then recovered after such a colossal stretch of time. A creature known as *Australopithecus garhi* lived in the Afar 2.5 million years ago. Its brain was only about the size of a large orange. Its muzzle jutted out below its eye sockets. It was superficially ape-like, but its canines were much smaller and blunter than any ape's, and its molars were much larger. Nearby we found evidence that large mammals had been butchered with stone tools, the earliest evidence of this kind of activity on Earth. The earliest stone tools, dated to 2.6 million years ago, also come from the Afar, about a day's drive along the camel tracks that stretch north from Bouri. North of the Middle Awash, a female's partial skeleton was trapped in sands at Hadar about 3.2 million years ago. She was found in 1974 by the Chicago paleoanthropologist Donald Johanson and nicknamed Lucy. Her species, *Australopithecus afarensis*, is well known from skeletal evidence at Hadar and from footprints in volcanic ash uncovered at Laetoli, in northern Tanzania. In the Middle Awash, we have found teeth, limb bones, and the most complete lower jaw of this species ever recovered.

The deeper you go into the paleoanthropological record, the more difficult it is to recover evidence. For years following the discovery and analysis of Lucy and her conspecifics, little was known about her ancestors. Only a few dental scraps of earlier hominids were found. But during the last decade, discoveries in several new fossil fields have allowed us to break through the 4-million-year barrier to find out about earlier hominids.

The Central Awash Complex is a block of sediments and interbedded lavas that dominates the landscape north of Bouri. Here in 1992, at a place called Aramis, our team probed a 4.4-million-year-old sediment package that yielded a primate called *Ardipithecus ramidus*, a creature like the younger *Australopithecus* but even more primitive. Five years later, Yohannes Haile-Selassie (the head of physical anthropology at the Cleveland Natural History Museum and no relation to the former emperor) began to recover primate teeth and bones from 5.7-million-year-old sediments in the foothills of the Middle Awash rift margin. This was an even more ape-like bipedal species, which he named *Ardipithecus kadabba*.

At this writing, we have recovered 227 hominid individuals, now under study at Ethiopia's National Museum, many of them already published in top science journals. So here, in one place in the Horn of Africa, we have an ordered set of snapshots through geological time. These glimpses, made available only during the past twenty-five years, are vital to understanding human origins and evolution, yet there are still many questions about our origins that we cannot answer, simply because we don't yet have enough fossils.

The Middle Awash sediment stack, with Herto man near the top, is but one set of windows on the past. Its unique record has allowed us to test those models that predict modern people all the way back to an allegedly abrupt origin in the oldest sediments. We have not found any evidence for that scenario. Rather, all the evidence is consistent with the hypothesis that the Herto man and his modern human descendants evolved. Each fossil our team collects from Middle Awash sediments tests hypotheses about evolution. We have now collected thousands of artifacts and 15,619 vertebrate individuals, from bats and shrews to crocodiles and gigantic elephants. We never find anatomically modern humans—or modern bats, or modern shrews, or modern hippos, or modern elephants—in the oldest sediments. We find modern species, like our own, in sediments at the top of the geological succession and extinct or ancestral ones at the bottom.

We have hundreds of fossils that anatomically and chronologically are intermediate between modern animals and older ancestors. Among the hominids, Herto is intermediate between living humans and the underlying Bodo man, and Bodo is intermediate between Herto and Daka. Given the 25,000 generations separating the Daka, Bodo, and Herto individuals, it is all the more amazing that Bodo is so anatomically and chronologically intermediate.

Will modern humans ever be found alongside *Australopithecus* in the older sediments at Afar or elsewhere? As scientists, we must leave that possibility open, but realistically it becomes more remote each time we find another fossil.

We can think of biological evolution as a singular experiment on a vast scale. The experiment recorded itself in the form of fossils and artifacts and their contexts—and also in the form of the modern biological world, a world rich in physical evidence of biological evolution. How do we go about reconstructing the experiment? The good news is that the experiment only happened once—there is only one past to reconstruct. The bad news is that our records of the experiment are so poor. But today's biological world is rich in readily available information. Let's start there.

Comparative biology shows a narrow genetic divide but broad behavioral and anatomical gulfs between modern humans and modern chimpanzees. The fossil record can illuminate from what sort of common ancestor those differences arose, and how. In the mid-1800s, the human fossil record was virtually nonexistent and even living African apes were little known; it would be over 100 years before the molecular anatomy of chimpanzees and humans could be compared. But the muscular and skeletal anatomy of humans and modern apes was available to early anatomists, like Thomas Henry Huxley. They performed careful comparative studies of bones, muscles, and organs of these primates and concluded that these apes were not our ancestors (after all, they were still alive) but rather our closest living relatives.

A century later, this conclusion would be amply confirmed by molecular studies. Today we know that the genomes of modern humans and modern chimpanzees differ by only a few percent. But compared to the living great apes—and all the hundreds of other primates (and thousands of other mammals, like rats and rabbits)—we humans are very odd. At first glance, we seem to stand apart. Why are we so bizarre, and how did we get this way? These are central issues in the study of human evolution.

Of the thousands of modern mammalian species produced by the grand experiment of evolution, there are only around 190 primate species. Of these, we humans are the most unusual. For this reason, people place themselves in their own zoological family, the Hominidae. Living humans are hominids with the following characteristics: We are the only primate (indeed, the only mammal) that walks habitually on two legs. Our brains are large relative to our body size. Our small canine teeth are set in tiny faces. Our technology is overwhelming, our communication system extraordinary. We tend to eat much meat, share our food, and display unique sexual anatomy and physiology. And we have traditionally interpreted ourselves as the pinnacle of the evolutionary process, the top of the tree. This view of the past is a narrow and misleading rendering of evolutionary history. The present has too often been imposed upon the past, the data of the past too quickly accommodated to preconceptions derived from the present. Our ancient human ancestors were neither humans nor chimpanzees—they were creatures busy being themselves. Some went extinct. Others became us.

We now understand that human ancestors were confined to Africa from the time when they split, sometime before 7 million years ago, from the line leading to living chimpanzees. The earliest hominids are found in wooded settings. Beginning about 4.2 million years ago, new hominids appeared in Africa, intermediate between the earlier ones and later ones, such as Lucy, but sharing with the latter the basic biology of a fundamentally bipedal locomotor pattern, ape-sized brains, and teeth and jaws much larger than ours. Lucy herself was neither chimp nor human. (Nor was she halfway between, because chimpanzees are still alive today.) Rather, Lucy was a member of a unique, widespread species well adapted to Pliocene Africa.

By 2.5 million years ago, a basic division among hominids had occurred, and specialized species of *Australopithecus* lived in southern and eastern Africa alongside subsequently smaller-toothed and larger-brained human ancestors who would expand from Africa. The earliest expansion happened some 1.8 million years ago, based on evidence from Georgia and Indonesia. These colonizers had spread across the Old World by the time the cycle of ice ages began, about a million years ago. By then, the specialized forms of *Australopithecus* had vanished from the African record.

European populations of the hominid colonizers would evolve into Neanderthals, an ice-adapted set of populations that evolved in accordance with the same evolutionary principles as other mammals, diverging from ancestral species and adapting to special conditions. Neanderthals, sophisticated bipeds who were much closer relatives of ours than any of the living apes are, went extinct long after the Herto people arose and spread from Africa. (The recently discovered remains of a dwarf hominid found in cave sediment on Flores Island in Indonesia may be evidence of yet another close-relative-recently-gone-extinct.)

We're all familiar with the iconography misleading us into imagining that the evolutionary experiment resulting in the modern world was an orderly, gradual, progressive affair, a ladder-like ascent to the present. But ever since Darwin, we have known this to be a narrow, self-serving view of the past. As Mark Twain observed in his essay "Was the World Made for Man?": "If the Eiffel tower were now representing the world's age, the skin of paint on the pinnacle-knob at its summit would represent man's share of that age; & anybody would perceive that that skin was what the tower was built for. I reckon they would. I dunno." Of course he knew.

Darwin's predictions about human evolution have been tested by the recovery of evidence showing that the living world was articulated by evolution. Humans were part of that process. There is a lot more to learn about how that happened. The ultimate cause(s) are currently intractable, but there's already plenty of evidence to show that evolution did happen and that we humans are among its products.

There are now tens of thousands of hominid fossils in museums around the world supporting our current knowledge of human evolution. The pattern that emerges from this vast body of hard evidence is consistent across thousands of investigations. All models, all myths involving the singular, instantaneous creation of modern humans fail in the face of this evidence. The origin stories developed a few thousand years ago by prehistoric pastoralists in the Middle East, hunter-gatherers in Australia, and farmers in Mesoamerica are all interesting and important, but these anthropological perspectives belong in the context of religious studies, not in science classrooms. And none of these myths have predicted what we have so far learned from the recovery and analysis of fossil evidence as well as from the biochemical and anatomical similarities and differences that characterize living things. A denial of evolution—however motivated—is a denial of evidence, a retreat from reason to ignorance.

Today, evolution is the bedrock of biology, from medicine to molecules, from AIDS to zebras. Biologists can't afford to ignore the interconnectedness of living things, much as politicians can't understand people, institutions, or countries without understanding their histories. As organisms, and as products of evolution, humans live and die in a biological world that has evolved and continues to evolve. The genes that determine much about how we live our lives—and how we die—are the products of evolution. Our lives depend on foods genetically modified by Neolithic ancestors. Our sufferings and deaths are often caused by microbes that have evolved resistance to our drugs. The implications of human evolution for law, medicine, agriculture, and biotechnology are vast. Our very futures are tied to emergent diseases and changing climates. We ignore these realities only at our peril.

The past is not a half-empty version of the present. An evolutionary perspective, based on the evidence, reveals past worlds as very different, complex, and fascinating places—places we still know very little about. Rather than characterizing evolution as a process designed to create the world for ourselves, our studies of evidence lead to understanding evolution as a vast ongoing experiment whose results happen to include us.

Nor is *Homo sapiens* the end product of evolution, the top rung on the ladder, the tip of the tree. We hominids were never a particularly bushy branch of that tree. As a species, we would never have gotten to where we are without our technology. Moreover, the evidence shows that our species is a peculiar creature whose cultural adaptation is currently perturbing the planet's ecosystems in drastic ways. Ironically, the world's lone remaining species of bipedal primate is poised to go down in evolutionary history as the greatest biological eliminator of species diversity that the planet has ever witnessed.

In this, the broader perspective, we are a fairly recent product of a still unfolding process. Human evolutionary history has important lessons for our species. The fossil record demonstrates that our closest relatives, bipedal primates far more like us than any modern ape, have repeatedly gone extinct during the last 6 million years. We can now recognize the outlines of our parentage in considerable detail: It is knowledge that should prove useful to a primate whose future is increasingly in its own hands.

TIM D. WHITE, a paleontologist at the University of California at Berkeley, is co-director of the Middle Awash project in the Afar Basin of Ethiopia, the largest and to date most successful multidisciplinary investigation into human origins and evolution. He is the author of *The Human Bone Manual.*

NEIL H. SHUBIN

The "Great" Transition

THE TAKE-HOME MESSAGE of this essay is a simple one: The transition of animals from water to land in the Devonian period, 370 million years ago, was profoundly important in one sense and entirely trivial in another. It had a major impact on our world, but it did not involve any unusual or extraordinary biological processes. The effects of the transition are all around us. We see them in the rocks. We see them in ponds and seas around the world. We even see them when we shake hands. Let me explain.

When we look back after 370 million years of evolution, the invasion of land by fish appears special. However, if we could transport ourselves by time machine to this early period, it isn't clear whether we would notice anything extraordinary. We would see a lot of fish, some of them big and some of them small, all of them struggling to survive and reproduce. Only now, 370 million years later, do we see that one of those fish sat at the base of a huge branch of the tree of life—a branch that includes everything from salamanders to humans. It would have taken an uncanny sixth sense for us to have predicted this outcome when our time machine deposited us in the middle of the Devonian.

To get a glimpse of the water-to-land transition, we need to see the creatures that lived on Earth at that time, then we need to look at our world today. When we do this, we see something sublime: The ancient world was transformed by ordinary mechanisms of evolution, with genes and biological processes that are still at work, both around us and inside our bodies.

The gulf between water and land looks like an unbridgeable divide. The challenges of life on land are vastly different from those in water. It would seem that completely different animals must live in these distinct habitats. Animals that walk on land need to cope with gravity; unlike water, air does not support animals as they move about. Animals also can dry out on land; this is particularly dangerous, because water is needed for many basic metabolic processes. And, of course, breathing is different in water than on land. Animals that breathe air need a more efficient way than gills to take in air and extract oxygen.

Because of all these factors, there are a daunting number of features that distinguish land-living animals from their fish ancestors: limbs with fingers and toes, necks, backbones with bony connections between vertebrae, a bony inner ear, a large scapula, ribs, paired nostrils, and so on. Biologists have singled out one of these characteristics for special treatment: True limbs are not seen in any living fish; for this reason, everything that is descended from fish is called a tetrapod (from the Greek for "four-footed").

For a long time it was thought that the shift from fish to tetrapod was driven by a transition from life in water to life on land. For example, it was thought that fins gradually evolved into limbs as animals began to walk. This thinking was captured by a famous hypothesis originally proposed by the American geologist Joseph Barrell in 1916 and later by the great American paleontologist Alfred Sherwood Romer. Romer and Barrell speculated that fish were forced out of water when Earth's climate supposedly became drier some 370 million years ago. As the ponds dried, so the story went, the fish had to learn to survive on land and so developed features that enabled them to hop from pond to pond.

When Romer did his work, in the 1920s through the 1960s, there was only one early tetrapod known: a limbed creature recovered from 365-million-year-old rocks in East Greenland. At present, East Greenland is a cold desert—dry, mountainous, and well above the Arctic Circle. Temperatures there rarely rise above freezing and for much of the year are colder than -20 F. But 365 million years ago East Greenland was a much warmer place, containing warm-water swamps, streams, and ponds. In the 1920s, a Swedish team led by Gunnar Save-Soderbergh discovered the skeletons of the then earliest known tetrapods in these rocks. These animals had robust limbs, appeared to be partly land-living, and supported Barrell's and Romer's hypothesis—at least, initially. To see how our theories have changed since Romer's day, we need to follow new evidence, whose trail leads to notions completely unforeseen even twenty years ago. This change in thinking attests to the power of evidence and the way it can change our view of the world.

In 1987 my colleague Jenny Clack began new studies in East Greenland and found the first important piece of evidence bearing on this water–land transition in over fifty years. She discovered the skeleton of another truly extraordinary tetrapod—one even more primitive than the one discovered by Save-Soderbergh. Sure enough, this creature has limbs with fingers and toes. It also has a very tetrapod-like hip, neck, and ear. What is remarkable is that this, the most primitive known tetrapod, is aquatic. It is not remotely specialized for life on land. It has fingers and toes but they are set within a limb that looks like a flipper. The limbs are delicate structures and seem unable to have supported the weight of the animal on land. It has a pair of hind limbs, but behind that is a tail that resembles that of a fish. Most important, this tetrapod has big gills.

The inescapable conclusion is that the most primitive tetrapod was an aquatic creature. The implications are profound: The fish-to-tetrapod transition likely happened not in creatures that were adapting to land but in creatures living in water. Moreover, everything special about tetrapods—limbs, digits, ribs, neck, the lot—might well have evolved in water, not on land.

This hypothesis made a prediction that could be tested: Aquatic animals more ancient than this new find should have intermediate structures. A search for these kinds of fossils dovetailed nicely with my own expeditionary research program in the late 1980s. Back then, my colleague Ted Daeschler and I were uncovering fish and tetrapods of the same age as Jenny's Greenland fossils in the roadcuts of central Pennsylvania. Pennsylvania is dotted with rocks of the same age as those of Greenland, but they need to be uncovered by dramatic means. The good news is that the state is not a frozen desert; the bad news is that fossils and rocks are mostly covered with trees, lawns, and cities. As a consequence, Ted and I made paleontological careers out of following the Pennsylvania Department of Transportation every time it cut a new road in central Pennsylvania. We found many fossils, but all of them were too young to test the issue at hand. We needed to go to a different area.

Ted and I ultimately found inspiration in an atypical place. We began a whole new research program that sprang from a single figure in a twenty-year-old textbook. I was thumbing through my old college geology text and found a map that seemed unremarkable at first. It was a map of North America with colored patches showing where rocks between 360 million and 380 million years old are preserved. One big splotch was on the east coast of Greenland, home to Jenny's find; another patch covered the part of Pennsylvania where most of our field effort had been focused. There was still another such area, though, and this is what made the figure interesting. Large, and running east–west across the Canadian Arctic, this patch extended over 500 miles and had never been explored by vertebrate paleontologists, although it was well known to geologists, particularly the Canadian geologists and paleobotanists who had mapped it extraordinarily well. The rocks turned out to be older than those in Pennsylvania and Greenland.

Ted and I first visited this area in 1999 and found little of interest. As it happened, we were fumbling around in the wrong part of the section; the rocks we were looking at were in the middle of an ancient ocean environment. When we shifted the expedition to areas that preserved ancient streams, lakes, and ponds, we found more fossils. During the 2004 field season, in these ancient environments, we found what we were looking for. Buried within a 370-million-year-old shallow stream was a collection of whole skeletons, one on top of the other. One of these creatures is an astonishing new kind of fish.

The new fish has fins, scales, and gills. By all definitions, it is a fish. This designation seems to hold until we look at its skeleton. Inside the fin is the skeletal pattern of all tetrapod limbs, in primitive form. It has an arm bone, a forearm, even a wrist. The new fish has a neck much like that of the earliest amphibians. The skull of this fish is not cone-shaped, as fish skulls are, but flattened like a crocodile's, with a nostril on either side. This creature also has expanded ribs, something unknown in any fish. We had found, one of my colleagues mentioned in jest, a fishapod.

The fishapod underscores one important point: It is no longer easy to distinguish a fish from a tetrapod. The Arctic fossils were only the tip of a paleontological iceberg. After subsequent discoveries in Latvia, Scotland, and China, the distinction is now so fuzzy that many of my colleagues do not even try to define tetrapods by ticking off a list of features. Our earlier definition of tetrapods distinguished them from fish by their possession of limbs. In what group, then, do we put our fish with wrists? What other characteristics might help us? Perhaps we could use lungs to distinguish tetrapods from fish. Then we would have to explain why lungfish use gills and lungs both, yet have fully formed fish fins. Scales? Even here, we run into the same problem, because early limbed and lunged animals also have belly scales. Indeed, the difficulty that our taxonomists have in distinguishing tetrapods from fish is the inevitable result of finding fossil intermediates.

This practical problem reflects a significant reality. One of the major transitions in the history of life is now bridged by a series of fossils dating from 380 million to 360 million years ago. The fact that we have discovered intermediates is not surprising; the surprise is that these creatures all appear to be aquatic and not specially adapted to life on land. This insight begs the question: Is there really a great divide between life in water and life on land? Answers to this question come from the study of fish alive today.

Modern fish have adapted to live in very different environments, including on the sea floor, in the shallows of lakes or streams, even partly in air. To cope with these environments, they have a remarkable set of features that enable them to walk, breathe, and even climb. For example, the various species of walking fish have evolved "armlike" bones and joints allowing them to prop up and propel their bodies along the ground. Some fish, like the mudskipper, maneuver in mudflats and spend a considerable period of their lives outside water, able to breathe air because the back of their mouth can absorb oxygen and relay it to the bloodstream. Mudskippers can hop good distances on the mudflats; some of them even climb trees by reaching up the trunk with their front fins and holding on with their hind fins.

What is important is that these various adaptations to land have evolved many times in fish. Several different kinds of fish climb trees; in addition, there are many different species of fish that breathe air, live part of their life on land, and walk about. The boundary between water and land is quite porous and bridged by modern fish from around the world. In fact, the adaptations we see in the fossils of the fish–tetrapod transition seem almost trivial in comparison to the living animals.

Mudskippers and the other walking fish are all very interesting, but are they extraordinary in an evolutionary sense? No, they are not, and the reason is instructive. Hopping, climbing, and breathing fish are just animals that have evolved to live in different kinds of aquatic and subaerial habitats. They are able to breathe air, hop, or climb because of subtle changes to their bodies; no revolutionary changes are needed. In evolutionary terms, the only way they will be notable is if their lineage is prolific and their descendants do great things. The relatives of the fish and tetrapods from Canada and Greenland were prolific; they are part of a trunk of the evolutionary tree that gave rise to every tetrapod—every bird, mammal, reptile, and amphibian. The mudskipper has a long way to go, and many hurdles to leap, before we will know whether its part of the evolutionary tree is special. If paleontologists 300 million years from now dig up the remains of a mudskipper, they will write chapters about its role in a "great" transition only if its part of the evolutionary tree has branched into many twigs. The mudskipper will get extra special treatment if one of its evolutionary branches leads to the paleontologists' own species.

Our understanding of the fish-to-tetrapod transition is not limited to long-dead fossils or obscure fish that climb trees. We have access to the DNA of every creature alive today. This is an enticing record of evolution, because DNA builds our bodies and is passed from generation to generation. By knowing how DNA works, we can dissect the molecular machinery that builds animals. This defines a whole new research program, one that was unimaginable in Romer's day. We can now compare the genetic recipe that builds a fish to the one that builds a tetrapod, in order to ask the question, What genetic changes are needed to turn a fish into a tetrapod? To see how this works, it helps to understand how DNA builds bodies. Every cell of our body has the same DNA inside. The various cells of our body are different because different genes are turned on and off in each cell. To understand what makes a cell in your eye different from a cell in the bones of your hand, we need to know about the genetic switches controlling the activity of genes in each venue. This leads us to the important point: These genetic switches help to assemble us. When we are conceived, we start as a single-celled embryo with the DNA needed to build our body. To go from this generalized cell to a complete human with trillions of specialized cells packed in just the right way, whole batteries of genes need to be turned on and off at just the right stages of development.

For evolutionists, this information is a boon. We can compare patterns of gene activity between different creatures to assess what kinds of changes are involved in the origin of new organs. Take appendages, for example. When we compare the ensemble of genes active in the development of a fish fin to those active in the development of a tetrapod limb, we can make a catalog of the genetic differences between fins and limbs. This comparison gives us some likely culprits—the genetic switches that may have changed during the origin of limbs. Based on what we know so far, the list is small: Very subtle changes in the activity of a relatively small number of genetic switches appear to underlie the differences between fins and limbs. To some extent, this should be obvious from the paleontological discoveries. Fins and limbs are part of a continuum, and no extraordinary events, processes, or genetic mechanisms are needed to explain the evolutionary transformation.

There are even clinical implications to all this. The genetic switches involved in the fin-to-limb transition are not 370-million-year-old relicts that lie in our bodies unchanged from generation to generation. Some of the genetic raw material of the fish-to-tetrapod transition still does business inside us. In fact, these genes continually mutate, sometimes with great consequences. Three hundred and seventy million years ago, changes to these genes led to the origin of limbs with fingers and toes. What happens when these genes change nowadays? Mutations can cause missing, malformed, or extra fingers in children.

We now know that the "great" transformation from water to land has so many fossil intermediates that we can no longer conveniently distinguish between fish and tetrapod, that living fish are bridging the water-to-land transition today, that some of the genes implicated in the ancient transition still reside and mutate in living animals, making everything from fish fins to human hands. Armed with this information, let's return to our opening handshake. The structures we shook with—our shoulder, elbow, and wrist—were first seen in fish living in streams over 370 million years ago. Our firm clasp is made with a modified fish fin. Actually, we carry an entire branch of the tree of life inside of us, and it does not stop there. That broad smile we give when we shake hands? The jaws that form our grin arose during another ancient "great" transition. The pair of eyes we use to make eye contact? These were the product of an even more ancient "great" transition. The list goes on and on. We can understand how all these things came about by using the same tools we did in this essay. Perhaps that is what is so profound about evolution: Everyday biological processes can explain things that seem special or mysterious about the living world. What is really powerful is that our explanations can be tested by an examination of the evidence.

The evolutionary biologist NEIL H. SHUBIN is chair of the Department of Organismal Biology and Anatomy at the University of Chicago.

RICHARD DAWKINS

Intelligent Aliens

AMONG THE MANY DISHONESTIES of the well-financed intelligent-design cabal is the pretense that the designer is not the God of Abraham but an intelligence unspecified, who could equally well be an extraterrestrial alien. Presumably the motive is to circumvent the First Amendment's prohibition on the establishment of religion, especially following Judge William Overton's 1982 decision in *McLean v. Arkansas Board of Education,* in which he struck down the state legislature's attempt to ensure "balanced treatment" in the schools for "creation science."

The religious affiliation of these people is not in doubt, and their in-group communications do not bother to hide their agenda. Jonathan Wells, one of the Discovery Institute's leading propagandists and the author of *Icons of Evolution,* is a lifelong member of the Unification Church (the Moonies). He wrote the following testimony in a Moonie in-house journal, under the heading "Darwinism: Why I Went for a Second PhD" (note that "Father" is the Moonies' name for Reverend Moon himself):

> Father's words, my studies, and my prayers convinced me that I should devote my life to destroying Darwinism, just

> as many of my fellow Unificationists had already devoted their lives to destroying Marxism. When Father chose me (along with about a dozen other seminary graduates) to enter a PhD program in 1978, I welcomed the opportunity to prepare myself for battle.

This quotation alone casts doubt on any claim Wells might have had to be taken seriously as a disinterested seeker after truth—which would seem a fairly minimal qualification for a PhD in science. He publicly admits to undertaking a scientific research degree not in order to discover something about the world but for the specific purpose of "destroying" a scientific idea that his religious leader opposed. Phillip Johnson, the born-again Christian law professor generally regarded as the leader of the gang, openly admits that his motive for opposing evolution is its "naturalism" (as opposed to supernaturalism).

The claim that the intelligent designer might be an alien from outer space may be disingenuous, but this doesn't stop it from serving as the basis for an interesting and revealing discussion. Such a constructive discussion, *within* science, is what I shall undertake in this essay.

The problem of recognizing an alien intelligence arises, in its starkest form, in that branch of science known as SETI, the Search for Extra-Terrestrial Intelligence. SETI deserves to be taken seriously. Its practitioners are not to be confused with those who complain of having been abducted in flying saucers for sexual purposes. For all sorts of reasons including the reach of our listening devices and the speed of light, it is extremely unlikely that our first apprehension of an alien intelligence will be a corporeal visitation. SETI scientists do not anticipate meeting extraterrestrial visitors in the flesh but in the form of radio transmissions whose intelligent origin should, it is hoped, be evident from their patterning.

A strong case can be made for the probable existence of intelligent life elsewhere in the universe. It gains support from the principle of mediocrity, that salutary lesson from Copernicus, Hubble, and others. Earth was once thought to be the only place in existence, surrounded by crystalline spheres bedecked with tiny stars. Later, when the size of the Milky Way galaxy was understood, it too was thought to be the only place, the locus of all that is. Then Edwin Hubble came along as a latter-day Copernicus to downgrade even our galaxy to mediocrity: It is only one among 100 billion galaxies in the universe. Today, cosmologists look at our universe and seriously speculate that it may be one of many universes in the "multiverse."

Similarly, the history of our species was once thought to have been roughly coterminous with the history of everything. Now, to borrow Mark Twain's crushing analogy, our history's proportionate duration has shrunk to the thickness of the paint on top of the Eiffel Tower. If we apply the principle of mediocrity to life on this planet, doesn't it warn us that we would be foolhardy and vain to think that Earth might be the only site of life in a universe of 100 billion galaxies?

It is a powerful argument, and I find myself persuaded by it. On the other hand, the principle of mediocrity is emasculated by another powerful principle, known as the anthropic principle, holding that the fact that we are in a position to observe the world's conditions determines that those conditions had to be favorable to our existence. The name comes from the British mathematician Brandon Carter, although he later preferred—with good reason—the "self-selection principle." I want to borrow Carter's principle for a discussion of the origin of life, the chemical event that forged the first self-replicating molecule and hence triggered natural selection of DNA and ultimately all of life. Suppose the origin of life really was a stupendously improbable event. Suppose the accident of primeval-soup chemistry which engendered the first self-replicating molecule was so prodigiously lucky that the odds against it were as low as one in a billion per billion planet years. Such fantastically low odds would mean that no chemist could entertain the smallest hope of repeating the event in a laboratory. The National Science Foundation would laugh in the face of a research proposal whose admitted chance of success was as low as one in a hundred per year, let alone one in a billion per billion years. Yet so great is the number of planets in the universe that even these minuscule odds yield an expectation that the universe contains a billion planets bearing life. And (here comes the anthropic principle) since we manifestly live here, Earth necessarily has to be one of the billion.

Even if the odds against life arising on a planet are as low as one in a billion billion (which puts it well beyond the range we would classify as possible), the plausible calculation that there are at least a billion billion planets in the universe provides an entirely satisfying explanation for our existence. There will still plausibly be one life-bearing planet in the universe. And once we have granted that, the anthropic principle does the rest. Any being contemplating the calculation necessarily has to be on that one life-bearing planet, which therefore has to be Earth.

This application of the anthropic principle is astonishing but watertight. I have oversimplified it by assuming that once life has originated on a planet, Darwinian natural selection will lead to intelligent and reflective beings. To be more precise, I should have been talking about the combined probability of life's originating on a planet and leading, eventually, to the evolution of intelligent beings capable of anthropic reflection. It could be that the chemical origin of a self-replicating molecule (the necessary trigger for the origin of natural selection) was a relatively probable event but later steps in the evolution of intelligent life were highly improbable. Mark Ridley, in *Mendel's Demon* (confusingly rebranded in America as *The Cooperative Gene*), suggests that the really improbable step in our kind of life was the origin of the eukaryotic cell. It follows from Ridley's argument that huge numbers of planets are home to something like bacterial life but only a tiny fraction of planets will have made it past the next hurdle to a level equivalent to the eukaryotic cell—what Ridley calls complex life. Or one might take the view that both those hurdles were relatively easy and that the really difficult step for terrestrial life was the attainment of the human level of intelligence. In this view, we would expect the universe to be rich in planets housing complex life but perhaps with only one planet harboring beings capable of noticing their own existence and therefore of invoking the anthropic principle. It doesn't matter how we distribute our odds among these three "hurdles" (or indeed other hurdles, such as the origin of a nervous system). So long as the total odds against a planet's evolving a life-form capable of anthropic reflection does not exceed the number of planets in the universe, we have an adequate and satisfying explanation for our existence.

Although this anthropic argument is entirely watertight, my strong intuitive feeling is that we do not need to invoke it. I suspect that the odds in favor of life's arising and subsequently evolving intelligence are sufficiently high that many billions of planets do indeed contain intelligent life-forms, many of them so superior to ours that we might be tempted to worship them as gods. Fortunately or unfortunately, we very likely won't encounter them: Even such apparently high estimates still leave intelligent life marooned on scattered islands, which might well be, on average, too far apart for their inhabitants ever to visit one another. Enrico Fermi's famous rhetorical question "Where are they?" could receive the disappointing answer "They are all over the place but too widely spaced to meet." Nevertheless, my belief, for what it's worth, is that the odds against intelligent life are nothing like as great as the anthropic calculation allows us to countenance. And therefore I think it is well worth putting quite a lot of money into SETI. A positive result would be an exhilarating biological finding, equaled in the history of biology perhaps only by Darwin's discovery of natural selection itself.

If SETI ever does pick up a signal, it will likely be from the high, or godlike, end of the spectrum of cosmic intelligences. We shall have a huge amount to learn from the aliens, especially about physics, which will be the same for them as it is for us. Biology will be very different, though—just how different will be a fascinating question. Communication will be all one-way. If Einstein is right about the limiting speed of light, dialogue will be impossible. We may learn from them, but we won't be able to tell them about us in return.

How, then, would we recognize intelligence in a pattern of radio waves picked up by a giant parabolic dish and known to originate from deep space and not a hoax? A tentative candidate was the pattern first detected by Jocelyn Bell Burnell in 1967 and jokingly called by her the LGM (Little Green Men) signal. This rhythmic pulse, with a periodicity of just over one second, is now known to have been a pulsar; indeed, hers was the first discovery of a pulsar. A pulsar is a neutron star rotating on its own axis with a beam of radio waves sweeping round and round as if from a lighthouse. That a star could rotate with a "day" measured on a scale of seconds is an extremely surprising fact—not the only surprising fact about neutron stars. But for present purposes the important point is that the periodicity of Bell Burnell's signal is not an indicator of intelligent origin but an unaided product of ordinary physics. Plenty of very simple physical phenomena, from dripping water to pendulums of all kinds, are capable of yielding rhythmic patterns.

What next might occur to a SETI researcher as diagnostic of intelligent life? Well, if we assume that the aliens actively want to signal their presence, we can ask what we would do if we were trying to transmit evidence of our intelligent presence. Certainly not emit a rhythmic pattern like Bell Burnell's LGM signal, but what else? Several people have suggested prime numbers as the simplest kind of signal that could originate only from an intelligent source. But how confident should we be that a pattern of pulses based on prime numbers could come only from a mathematically sophisticated civilization? Strictly speaking, you can't prove that there is no inanimate physical system capable of generating prime numbers. You can say only that no physicist has ever discovered a non-biological process capable of generating them. Strictly speaking, the same caution goes for any signal. However, there are certain kinds of signals—of which those based on prime numbers may be the simplest example—which would be so convincing as to leave alternatives looking absurd.

Disquietingly, biologists have proposed models that are capable of generating prime numbers but do not involve intelligent design. Periodical cicadas emerge for breeding every seventeen years (in some varieties) or every thirteen years (in other varieties). Two theories to account for this odd periodicity depend on the fact that 13 and 17 are prime numbers. I'll describe just one of these theories. Its premise is that plague-year breeding is an adaptation to foil predators by swamping them. But then predator species evolved their own periodic breeding pattern to cash in on cicada plagues (or bonanzas, as they would see it). In an evolutionary arms race, the cicadas "replied" by lengthening the period between plague years. The predators lengthened theirs in response. (Remember that this shorthand language of "reply" and "response" implies no conscious decisions, only blind natural selection.) When in the course of the arms race the cicadas reached an interval, such as six years, which was divisible by some other number, the predators found it more profitable to drop their breeding interval, to, say, three years, thereby hitting the cicada bonanza with alternate peaks of their own breeding cycle. Only when the cicadas hit a prime number did this become impossible. The cicadas continued to lengthen their stride until they reached a number that was too large to allow the predators to synchronize directly, yet prime and therefore impossible to meet with some multiple of a smaller period.

Well, that may not seem a very plausible theory, but it doesn't need to be for my purpose. I simply need to show that it is possible to conceive of a mechanistic model that does not involve conscious mathematics yet still manages to generate prime numbers. The cicada example shows that while prime numbers may not be generatable by non-biological physics, they can be generated by non-intelligent biology. Even the implausible cicada story is a cautionary tale to warn us that at least it is not necessarily obvious that prime numbers are diagnostic of intelligence.

The difficulty of diagnosing intelligence in a radio signal is itself a cautionary tale that calls to mind the historical analogy of the argument from design. There was a time when everybody (with very few, very distinguished exceptions, such as David Hume) thought it completely obvious that the complexity of life was unmistakably diagnostic of intelligent design. What should give us pause is this: Darwin's nineteenth-century contemporaries could claim the right to be as surprised by his remarkable discovery as we should be surprised today if a physicist discovered an inanimate mechanism capable of generating prime numbers. Perhaps we should entertain the possibility that other principles, comparable to Darwin's, remain to be discovered—principles capable of mimicking an illusion of design as convincing as the illusion manufactured by natural selection.

I am not inclined to predict any such event. Natural selection itself, properly understood, is powerful enough to generate complexity and the illusion of design to an almost limitless extent. Bear in mind that elsewhere in the universe there could be variants of natural selection that, although based on essentially the same principle as Darwin discovered on this planet, might be almost unrecognizably different in detail. Bear in mind, too, that natural selection can midwife other forms of design. It doesn't stop with its direct productions, such as feathers, ears, and brains. Once natural selection has produced brains (or some extraterrestrial equivalent of brains), those brains can go on to produce technology (extraterrestrial equivalents of technology), especially computers (extraterrestrial equivalents of them, too). The manifestations of deliberate engineering design—indirect rather than direct productions of natural selection—can burgeon into new reaches of complexity and elegance. The point here is that natural selection manifests itself in the form of design at two levels: There is, first, the *illusion* of design, which we see in a bird's wing or a human eye or brain; and, second, there is "true" design, which is a product of evolved brains.

And now to my central point. There really is a profound difference between an intelligent designer who is the product of a long period of evolution, whether on this planet or a distant one, and an intelligent designer who just *happened*, without any evolutionary history. When a creationist says that an eye or a bacterial flagellum or a blood-clotting mechanism is so complex that it must have been designed, it makes all the difference in the world whether the "designer" is thought to be an alien produced by gradual evolution on a distant planet or a supernatural god who didn't evolve. Gradual evolution is a genuine explanation, which really can theoretically yield an intelligence of sufficient complexity to design machines and other things too complex to have come about by any process other than design. Hypothetical "designers" jumped-up from nothing cannot explain anything, because they can't explain themselves.

There are some man-made machines that common sense, if not strict logic, tells us could not have come about by any process other than intelligent design. A jet fighter, a moon rocket, a motorcar, a bicycle—these are surely designed. But the important point is that the entity that did the designing—the human brain—is not. There is overwhelming evidence that the human brain evolved through a graded series of almost imperceptibly improving intermediates, whose relics may be seen in the fossil record and whose analogs survive all around the animal kingdom. Moreover, Darwin and his twentieth- and twenty-first-century successors have provided us with a luminously plausible explanation for the mechanism that propels evolution up the graded slopes, the process I have dubbed "Climbing Mount Improbable." Natural selection is not some desperate last resort of a theory. It is an idea whose plausibility and power hits you between the eyes with a stunning force, once you understand it in all its elegant simplicity. Well might T. H. Huxley cry out, "How extremely stupid not to have thought of that!"

But we can go further. Not only does natural selection explain the bacterial flagellum, the eye, the feather, and brains capable of intelligent design. Not only can it explain every biological phenomenon ever described. It is the only plausible explanation for these things that has ever been proposed. Above all, the argument from improbability—the very argument that the advocates of intelligent design fondly imagine supports their case—turns around to kick that case over with devastating force and lethal effect.

The argument from improbability states, incontrovertibly, that some phenomenon in nature—something like a bacterial flagellum, say, or an eye—is too improbable to have simply happened. It has to be the product of some very special process that generates improbability. The mistake is to jump to the conclusion that "design" is that very special process. In fact it is natural selection. The late Sir Fred Hoyle's jocular analogy of the Boeing 747 is useful, although it, too, turns out to make the opposite point to the one he intended. The spontaneous origin of the complexity of life, he said, is as improbable as a hurricane blowing through a scrap yard and spontaneously assembling a Boeing 747.

Everybody agrees that airliners and living bodies are too improbable to be assembled by chance. A more precise characterization of the kind of improbability we are talking about is *specified improbability* (or specified complexity). The "specified" is important, for reasons that I explained in *The Blind Watchmaker* (1986). I first pointed out that randomly hitting the number that opens the large combination lock on a bank vault is improbable in the same sense as hurling scrap metal around and assembling an airliner:

> Of all the millions of unique and, with hindsight equally improbable, positions of the combination lock, only one opens the lock. Similarly, of all the millions of unique and, with hindsight equally improbable arrangements of a heap of junk, only one (or very few) will fly. The uniqueness of the arrangement that flies, or that opens the safe, is nothing to do with hindsight. It is specified in advance. The lock-manufacturer fixed the combination, and he has told the bank manager. The ability to fly is a property of an airliner that we specify in advance. . . . (p. 10)

Given that chance is ruled out for sufficient levels of improbability, we know of only two processes that can generate specified improbability. They are intelligent design and natural selection, and only the latter is capable of serving as an ultimate explanation. It generates specified improbability from a starting point of great simplicity. Intelligent design can't do that, because the designer must itself be an entity at an extremely high level of specified improbability. Whereas the specification of the Boeing 747 is that it must be able to fly, the specification of the "intelligent designer" is that it must be able to design. And intelligent design cannot be the ultimate explanation for anything, for it begs the question of its own origin.

From the lowlands of primeval simplicity, natural selection gradually and steadily ramps its way up the gentle slopes of Mount Improbable until, after sufficient geological time, the end product of evolution is an object such as an eye or a heart—something of such an elevated level of specified improbability that no sane person could attribute it to random chance. The single most unfortunate misunderstanding of Darwinism is that it is a theory of chance; the misunderstanding presumably stems from the fact that mutation is random. But natural selection is anything but random. To escape from chance is the primary accomplishment that any theory of life must aspire to. Obviously, if natural selection were a theory of random chance, it could not be right. Darwinian natural selection is the *non*-random survival of randomly varying coded instructions for building bodies.

Engineers use explicitly Darwinian methods in the computer simulations they make in order to optimize systems. They escalate performance from poor beginnings up a ramp of improvement to something approaching an optimum. The engineer's wastepaper basket holds the "mutant" designs he discarded before putting them to the test. Some designs don't even make it onto paper but are discarded in the engineer's head. I have no need to pursue the question of whether Darwinian natural selection is a good or helpful model for what goes on in the brain of a creative engineer or artist; constructive creative work—by engineers or artists, or indeed anybody—may or may not plausibly represent a form of Darwinism. The fundamental point remains that all specified complexity must ultimately rise from simplicity by some kind of escalatory process. If we ever discover evidence that some aspect of life on Earth is so complex that it must have been designed, scientists will face with equanimity—and doubtless some excitement—the possibility that it was designed by an extraterrestrial intelligence. The molecular biologist Francis Crick, together with his colleague Leslie Orgel, made such a suggestion in proposing the theory of directed panspermia. According to Orgel and Crick's idea, extraterrestrial designers deliberately seeded Earth with bacterial life. But the important point is that the designers were themselves the end product of some extraterrestrial version of Darwinian natural selection. The supernatural explanation fails to explain because it ducks the responsibility to explain itself.

Creationists who disguise themselves as "intelligent-design theorists" have only one argument, and it goes like this:

1. The eye (the articulation of the mammalian jaw, the bacterial flagellum, the elbow joint of the lesser spotted weasel frog—which you have never heard of and don't have time to look up before you seem to a lay audience to have lost the argument) is irreducibly complex.
2. Therefore it cannot have evolved by gradual degrees.
3. Therefore it must have been designed.

No supporting evidence is ever offered for step 1, the allegation of irreducible complexity. I have sometimes referred to it as the argument from personal incredulity. It is always offered as a negative argument: Theory A is alleged to fail in some respect, so we have to default to theory B, without even asking whether theory B might be deficient in the very same respect.

One legitimate response of biologists to the argument from personal incredulity is to attack step 2: Look carefully at the examples proposed and show that they either did, or could, easily evolve by gradual degrees. Darwin did this for the eye. Later paleontologists did it for the articulation of the mammalian jaw. Modern biochemists have done it for the bacterial flagellum.

But the message of this essay is that, strictly speaking, we needn't bother to dispute steps 1 and 2. Even if they were ever accepted, step 3 remains irretrievably invalid. If incontrovertible evidence of intelligent design were ever discovered in, say, the organization of the bacterial cell—if we found evidence as strong as a manufacturer's signature written in unmistakable characters of DNA—this could only be evidence of a designer that was itself the product of natural selection or of some other as yet unknown escalatory process. If such evidence were found, our minds should immediately start working along the lines of Crick's directed panspermia, not a supernatural designer. Whatever else irreducible complexity might demonstrate, the one thing it cannot appeal to in ultimate explanation is something else that is irreducibly complex. Either you accept the argument from improbability, in which case it disproves the existence of ultimate designers. Or you don't accept it, in which case any attempt to deploy it against evolution is inconsistent if not dishonest. You cannot have it both ways.

RICHARD DAWKINS, an evolutionary biologist, is Charles Simonyi professor of the public understanding of science at Oxford University. His latest book is *The Ancestor's Tale: A Pilgrimage to the Dawn of Life*.

FRANK J. SULLOWAY

Why Darwin Rejected Intelligent Design

THERE IS CONSIDERABLE IRONY in the fact that Charles Darwin was at one time enthralled by the theory that all species are intelligently designed—a theory he later sought to banish from science in his *Origin of Species* (1859). Popularized in the seventeenth century in works such as John Ray's *The Wisdom of God Manifested in the Works of the Creation* (1691), this doctrine sought to unite a celebration of God's handiwork in the Creation with the pursuit of natural science. Such theologically inspired arguments reached their apogee a century later in the writings of William Paley. A clergyman like Ray, Paley set out his influential ideas in *Natural Theology; or, Evidences of the Existence and Attributes of the Deity, Collected from the Appearances of Nature* (1802). The many proofs he adduced in favor of intelligent design, including such examples as the hinges on bivalve shells and the plumes that facilitate wind dispersal of certain seeds, fascinated and convinced young Darwin.

Natural Theology is constructed around the analogy of a watchmaker, a metaphor borrowed from previous advocates of the design doctrine. Imagine, Paley suggests, that you are walking across a heath and suddenly encounter a watch lying on the ground. After close inspection of the watch, you would be compelled to conclude that such an intricate device could not have been constructed otherwise in order for it to work. It is only reasonable to assume "that the watch must have had a maker: that there must have existed, at some time and at some place or other, an artificer or artificers who formed it for the purpose which we find it actually to answer, who comprehended its construction and designed its use." In the case of living organisms, Paley continues, the evidence for design is even stronger, "in a degree which exceeds all computation," and he concludes: "The marks of *design* are too strong to be gotten over. Design must have had a designer. That designer must have been a person. That person is GOD."

The efforts by Ray, Paley, and others to unite natural history with theology were among the inducements that inclined Darwin, at the age of eighteen, to look favorably on a career in the church. Sent by his father two years earlier to study medicine at Edinburgh University, Darwin had found himself uninspired by this profession and revolted by the sight of operations, which at that time were conducted without the benefit of anesthesia. Worried that his son might turn into "an idle sporting man," Darwin's father, a successful physician, proposed that he attend Cambridge University in preparation for becoming an ordained minister. "Considering how fiercely I have been attacked by the orthodox," Darwin later reflected on this youthful career decision in his 1876 *Autobiography,* "it seems ludicrous that I once intended to be a clergyman."

At Christ's College, Cambridge, Darwin was assigned the same rooms that Paley had occupied more than a half century earlier. Paley's *Evidences of Christianity* (1794) and his *Principles of Moral and Political Philosophy* (1785) were among the books selected as examination topics for Darwin's BA degree in 1830. Darwin studied these works diligently, as he later recalled in his *Autobiography*:

> I am convinced that I could have written out the whole of the *Evidences* with perfect correctness, but not of course in the clear language of Paley. The logic of this book and as I may add of his *Natural Theology*, gave me as much delight as did Euclid. The careful study of these works, without attempting to learn any part by rote, was the only part of the Academical Course which, as I then felt and as I still believe, was of the least use to me in the education of my mind. I did not at that time trouble myself about Paley's premises; and taking these on trust I was charmed and convinced by the long line of argumentation.

On the very day of publication of the *Origin of Species* three decades later, Darwin reiterated his praise for Paley in a letter to his neighbor in Downe, John Lubbock: "I do not think I hardly ever admired a book more than Paley's *Natural Theology*: I could almost formerly have said it by heart." Even after the *Origin* had finally turned the tide in thinking about the theory of evolution and had introduced Darwin's even more controversial mechanism of evolutionary change—natural selection—Darwin was still subject to the compelling lure of the design doctrine. The Duke of Argyll reported a conversation he had had with Darwin on this issue during the last year of Darwin's life. The Duke had maintained that he could not look upon certain adaptations, such as those found in orchids, "without seeing that they were the effect and the expression of Mind. I shall never forget Mr. Darwin's answer. He looked at me very hard and said, 'Well, that often comes over me with overwhelming force; but at other times,' and he shook his head vaguely, adding, 'it seems to go away.' "

What triggered Darwin's dramatic change of mind about the origin of species was his five-year voyage around the world on HMS *Beagle*, and especially his five-week visit to the Galápagos Islands in September and October 1835. Legend has it that Darwin underwent a eureka-like conversion to the theory of evolution during this brief visit. Shorn of the legend, the actual story of Darwin's conversion, which did not occur until after his return to England a year and a half later, tells us far more about how science is really done, especially how theory guides observation and prepares the mind, and how dogged persistence is required to transform controversial theories into widely accepted facts.

While in the Galápagos, creationist theory primed Darwin in key ways for what he observed and understood there. Just as important, this theory also dictated what he failed to observe and understand. Commenting on his extensive efforts to collect specimens on Charles Island, the second of the four islands he visited, Darwin recorded in his personal journal, "It will be very interesting to find from future comparison to what district or 'centre of creation' the organized beings of this archipelago must be attached." He was clearly trying to reconcile the new and strange creatures he was encountering in this remote archipelago with the prevailing creationist paradigm. According to this theory, different "centers of creation" explained why the earth's flora and fauna differed from one region to another—for example, between continents. What is apparent from Darwin's reference is that he did not yet realize that such a tiny portion of the globe as the Galápagos archipelago might actually be its own "center of creation."

Today, as a result of Darwin's change of scientific heart, the Galápagos are widely recognized as a classic laboratory of evolution in action, and hence (to use the pre-Darwinian terminology) as a unique center of creation in their own right. In significant part, the islands owe their biological celebrity to the fact that they are geologically young (about 3 million years old, although former islands, now eroded down and submerged beneath the sea, date back as much as 17 million years).[1] Based on his own geological observations, Darwin correctly realized that life in these distant islands had been given a new beginning. As he later reflected on this insight in his *Journal of Researches*:

> Seeing every height crowned with its crater, and the boundaries of most of the lava-streams still distinct, we are led to believe that within a period geologically recent the unbroken ocean was here spread out. Hence, both in space and time, we seem to be brought somewhat near to that great fact—that mystery of mysteries—the first appearance of new beings on this earth.

Certain unexpected facts about the Galápagos undermined the credibility of any creationist solution to this "mystery of mysteries." In particular, each island in the Galápagos group has evolved many of its own distinct species over time, following the stocking of the islands by chance colonists who managed to arrive there from the Central and South American mainland, almost 600 miles away. Darwin was first alerted to this possibility by Nicholas Lawson, the vice-governor of the islands, whom Darwin

---

[1] See D. M. Christie et al., "Drowned islands downstream from the Galápagos hotspot imply extended speciation times," *Nature* 355 (1992):246–48; and R. Werner and K. Hoernle, "New volcanological and volatile data provide strong support for the continuous existence of Galápagos Islands over the past 17 million years," *Int. J. Earth Sci.* 92 (2003):904–11.

encountered on Charles Island. As he reports in his *Journal of Researches*, Lawson insisted that "the tortoises differed from the different islands, and that he could with certainty tell from which island any one was brought." Unfortunately, Darwin did not at first pay sufficient attention to the vice-governor's testimony, in part because of an error in the contemporary zoological literature. Like other naturalists, Darwin was under the mistaken impression that the Galápagos tortoise, which was at that time classified under the misleading scientific name *Testudo indicus*, was not native to the Galápagos but had been transported there by buccaneers or perhaps oceangoing Polynesians, from islands in the Indian Ocean, where similar forms of giant tortoise are found. Only after Darwin's return to England did it become clear that these two were separate species.

There was a second reason why Darwin initially overlooked the vice-governor's remarks about the tortoises. Creationist theory held that species can and do change in response to local environments. Like an elastic band that resists being stretched, any departure among varieties from the supposedly immutable specific type was thought to be a temporary deviation, maintained by unusual ecological conditions. Darwin therefore appears to have concluded, during his Galápagos visit, that possible island-to-island differences among the giant land tortoises were no more remarkable than if introduced goats were also to differ from island to island in characteristics like color and size. Such local differences were readily interpretable as short-term perturbations somehow induced by transport to new and demanding environments.

As a result of his creationist perspective on species and varieties, Darwin—astonishingly, from our modern evolutionary viewpoint—failed to collect a single specimen of giant tortoise for scientific purposes during his Galápagos stay. After he had left the archipelago and was sailing to Tahiti, Darwin had one last opportunity to rectify this collecting oversight. The *Beagle* had stocked forty-eight tortoises from San Cristóbal, the first of the four islands Darwin had visited. As they sailed across the Pacific, Darwin and his fellow shipmates gradually ate their way through the evidence that was later, in the form of hearsay, to revolutionize the history of science. The carapaces of those forty-eight tortoises, which could have been compared with other specimens at European museums, were unfortunately all thrown overboard with the rest of the *Beagle*'s garbage.

This same creationist mindset helps to explain why Darwin at first failed to understand the most famous Galápagos exemplar of evolution in action—namely, Darwin's famous finches. Fourteen species of finches have evolved in the Galápagos Islands, from an ancestral form now known to be a member of the *Tiara* group (grassquits), which is found in Central and South America.[2] Over the last 2 million years or so, this evolutionary process has resulted in such impressive adaptive radiation into diverse ecological niches that some of these fourteen Galápagos species are not particularly finchlike in appearance. Although four of the fourteen species feed on seeds, as finches generally do, another two species feed on the fruits, flesh, and flowers of cactus. Seven additional species are primarily insectivorous, and one very remarkable species feeds almost exclusively on leaves. It is not surprising, then, that Darwin was fooled by some of these finches into thinking they were not finches at all but the species that they had come to mimic through a process known as convergent evolution. In his specimen notebook,

---

[2] See K. Petren, B. R. Grant, & P. R. Grant, "A phylogeny of Darwin's finches based on microsatellite DNA length variation," *Proc. Roy. Soc. Lond. B* 266 (1999):321–29.

for example, he listed the warbler finch as a "Wren"—which is what this bird, to an untrained eye, appears to be. Similarly, he listed the small cactus finch, which has a long pointed bill for getting at the fruits and flowers of the *Opuntia* cactus, as a member of the Icteridae, which is the family of meadowlarks and orioles.

So confusing was the case of the Galápagos finches that Darwin did not grasp, at the time of their collection, that all these birds were closely related or that the peculiarly large number of species in this one avian group might result from their having evolved on different islands. He therefore made no effort to label his ornithological collections by island, something he was later to regret when he finally surmised the evolutionary origins of this peculiar group of birds. Nor did Darwin have the opportunity to observe these finches in sufficient detail to realize that their beak sizes and shapes were closely related to their diets, an important insight that the legend wrongly ascribes to him.

Despite being armed with an inadequate theory during his Galápagos visit—a theory that served to undermine his collecting methodology by focusing his attention on the Galápagos as part of a presumably larger center of creation, Darwin was too good a naturalist not to notice that the four mockingbird specimens he collected, each from a different island, were either distinct varieties or separate species. Not being an expert in ornithology, Darwin was unsure just what to make of this anomaly, but he did record localities for these four specimens. In July 1836, nine months after his Galápagos visit, Darwin reflected on this unusual case of the mockingbirds and recalled as well what he had been told about the tortoises:

> When I see these Islands in sight of each other, & possessed of but a scanty stock of animals, tenanted by these birds but

> slightly differing in structure & filling the same place in Nature, I must suspect they are only varieties. . . . If there is the slightest foundation for these remarks the zoology of Archipelagoes—will be well worth examining; for such facts would undermine the stability of Species.

The key to interpreting this famous passage—which broaches the Darwinian revolution only to back away from it—is the phrase "I must suspect they are only varieties," a presumption that Darwin understood to be fully consistent with creationist theory. What kept Darwin from taking the crucial step from scientific orthodoxy to heterodoxy at this point in the *Beagle* voyage was a lack of critical information about proper ornithological classification, which would only become available to him back in England through access to museum collections and especially the judgments of expert ornithologists, who were much more familiar than Darwin was with the defining features of species and varieties in the particular Galápagos taxa he had collected. Faced with an absence of crucial information to resolve the issue, Darwin continued to give a nod toward the prevailing creationist assumption that variation within immutable species can lead to new varieties or subspecies that are adapted to local environments. Despite any doubts Darwin may have entertained on this subject during the remainder of the *Beagle* voyage, this scientific position was perhaps the most responsible one for him to have taken at that time.

Darwin returned to England on October 2, 1836. Three months later, he deposited his *Beagle* collections of birds with John Gould, the ornithological expert at the London Zoological Society. Gould, who was in the process of becoming famous for his beautifully illustrated monographs on birds of the world, immediately realized the extraordinary nature of Darwin's Galápagos specimens and analyzed and described them first, ahead of the other birds from the *Beagle* voyage.

Darwin did not obtain a full report on Gould's findings until early March 1837, when he moved from Cambridge to London. After the two men had discussed these findings in detail, Darwin's life and scientific thinking were never the same. Gould informed Darwin that three of his four specimens of Galápagos mockingbird were distinct species, new to science and different from all known mockingbirds. Gould also informed Darwin that his collection included thirteen or possibly fourteen species of very unusual finches, all so closely related that Gould had placed them in a single new group. As for the land birds as a whole, twenty-five out of twenty-six were judged to be new to ornithology and unique to the Galápagos, something Darwin could not have known without access to the museum collections and previously published descriptions available to Gould. All of a sudden, following Gould's taxonomic analyses, the Galápagos Islands had become their own distinct "center of creation." Darwin now found himself confronted by the problem of the origin of species in a way he had not when he was visiting those islands.

Darwin appears to have been stunned by Gould's conclusions. He quickly realized that if Gould was right about the mockingbirds (and Darwin apparently pressed him on this point, to be absolutely certain), the supposedly immutable barrier that exists between separate species had somehow been broken by these birds, isolated on the different islands of the Galápagos group. Gradual evolution through geographic isolation was the only plausible explanation, unless one believed that God, like an obsessive-compulsive gardener, had gone from one island to the next in the Galápagos group whimsically creating separate but closely allied species intended to fill the same ecological niches. As for the much more complex case of the finches, once Gould had convinced Darwin that these birds were all closely related, based on the simpler and more compelling case of the mockingbirds, Darwin was able to see the finches in a radically new light. As he later observed in his *Journal of Researches*: "Seeing this gradation and diversity of structure in one small, intimately related group of birds, one might really fancy that from an original paucity of birds in this archipelago, one species had been taken and modified for different ends." Only now did he appreciate the extent of his previous oversight in failing to label the bulk of his Galápagos birds by island. Such evidence, he realized, would help to account for why so many different species of finches lived there.

Fortunately, Darwin knew that three other collectors on the *Beagle* (Captain FitzRoy, FitzRoy's steward Harry Fuller, and Darwin's own servant, Syms Covington) had also collected specimens in the Galápagos. All of these specimens turned out to have been labeled by island; significantly, it was the nonscientists on the *Beagle*, who were not as theory-driven as Darwin, who recorded the scientific evidence that Darwin, based on a creationist approach, had considered superfluous. After his meeting with Gould, Darwin diligently sought out this locality information, and he later used it to support his case about the mockingbirds and tortoises, although the evidence was still uncomfortably tenuous. Fortunately, additional sources of information over the next two decades, drawn in part from reports about other oceanic archipelagoes, would transform this suggestive evidence into incontrovertible fact.

Jolted by the Galápagos results, and struck by similar evidence about geographic distribution that began to emerge from taxonomic judgments about his South American collections, Darwin began a search in the spring of 1837 for a mechanism that would explain not only evolutionary change but also the most elusive puzzle about species—their highly adaptive nature. As he later recorded in a private journal: "In July [1837] opened first note book on 'Transmutation of Species'—Had been greatly struck from about Month of previous March on character of S. American fossils—& species on Galapagos Archipelago. These facts origin (especially latter) of all my views."

Based on the evidence of the Galápagos mockingbirds, Darwin now understood that geographic isolation was a crucial part of the answer to how species transform themselves over time. But isolation, although it could account for the origin and multiplication of new species over time, was insufficient to explain the often remarkable adaptations that species manifest to local environments. After exploring and rejecting a number of hypotheses about evolutionary change, Darwin happened to read, in September 1838, the 1826 edition of Thomas Malthus's *Essay on the Principle of Population*. Malthus argued that populations have an inherent tendency to grow geometrically. Yet, in nature, the food supply is limited, so most offspring do not survive, being killed by predators, famine, and diseases. On reading Malthus's book, Darwin immediately realized that in the ever-present struggle for existence, slight variations of a beneficial nature would tend to be naturally selected, leading to increased survival and hence an increase in adaptive traits, just as the breeder of domesticated animals achieves desired traits by selecting the qualities that are esteemed in these animals. "Here, then," Darwin remarked in his *Autobiography*, "I had at last got a theory by which to work." Here also was a credible answer to William Paley—one entailing a natural (as opposed to supernatural) explanation of adaptations. Natural selection, Darwin realized, was none other than Paley's designer, an equivalence that later led Darwin sometimes to personify this evolutionary process, as in this celebrated passage from the *Origin*: "It may be said that natural selection is daily and hourly scrutinising, throughout the world, every variation, even the slightest; rejecting that which is bad, preserving and adding up all that is good; silently and insensibly working, whenever and wherever opportunity offers, at the improvement of each organic being in relation to its organic and inorganic conditions of life."

Inspired by the striking evidence from the Galápagos Islands, and armed with his novel theory of natural selection, Darwin began to reexamine the basic assumptions of creationism and to compare the predictions one would make based on these two radically different theories. The more extensive his reexamination became, the more he realized that the theory of intelligent design, which gave creationism its scientific legitimacy, was overwhelmingly contradicted by the available evidence. Darwin's reassessment reached its culmination twenty-two years later in *On the Origin of Species by Means of Natural Selection*, a book that Darwin himself aptly characterized toward the end as "one long argument." It was as much an argument *against* creationism, and especially against the validity of intelligent design, as it was an argument *for* evolution, as Ernst Mayr has noted in his 1991 book, *One Long Argument: Charles Darwin and the Genesis of Modern Evolutionary Thought*. Although Paley is mentioned only once in the *Origin*, his specter is everywhere apparent in Darwin's repeated arguments highlighting the lack of truly intelligent design in nature.

Not surprisingly, the evidence about geographical distribution, particularly about oceanic islands and their biological relationships with the nearest continents, plays a substantial role in Darwin's argument. The numerous problems that such islands raised for creationism had been surprisingly overlooked prior to Darwin's discussions of the subject, and this powerful class of evidence was the only topic to which Darwin devoted two whole chapters in the *Origin*. The Galápagos, for example, are home to various species of animals and plants closely allied to those from the neighboring American continent, yet the environmental features of these islands do not at all resemble those of the nearest parts of the continent, which are tropical. By contrast, the harsh volcanic environment of the Galápagos does closely resemble that of the Cape Verde Islands, 400 miles west of Africa. Yet the Cape Verdean flora and fauna are most strongly allied to species living on the African mainland, not to those in the Galápagos. Why would a Designer, Darwin asked, place two completely different creative stamps—one African and one American—on species that live in nearly identical environments and fill similar ecological niches? Creationist theory, he argued, ought to predict that such island species would either be identical or closely allied, based on the similar environments to which they are supposedly adapted by intelligent design. But the true circumstances are "utterly inexplicable on the ordinary view of independent creation of each species." By contrast, anyone who accepts the theory of evolution would expect precisely this sort of evidence.

In the *Origin*, Darwin extended this general argument about oceanic islands to include so-called relict populations on continents, such as the blind organisms (rodents, reptiles, insects, crustaceans, and fish) that live within deep caverns. Almost no environments, Darwin argued, could be more similar than the habitats that characterize such cavern systems, because these caverns are insulated from major changes in temperature, humidity, and the seasons. Yet blind cave animals, which often have rudimentary eyes that serve no purpose, are not closely allied to one another around the world. Instead, these organisms most closely resemble the surface-dwelling organisms on each of the continents where the caverns are found and from which their denizens have been derived by evolution. In short, such facts about relict populations do not admit of a rational explanation in terms of the theories of independent creation and intelligent design. Such facts do, however, agree with the expectations derived from evolutionary theory.

Islands like the Galápagos focused Darwin's attention on another class of facts inimical to the theory of intelligent design. Remote oceanic islands have very skewed faunal distributions. Absent from such island populations are large terrestrial mammals as well as amphibians (frogs, toads, and newts). Yet bats, which can reach oceanic islands by flight, are regularly found to inhabit the islands of every major ocean. The physical conditions of such islands, Darwin maintained, cannot explain the absence of amphibians and large terrestrial mammals, for when such animals have been introduced to oceanic islands—and Darwin presented relevant evidence in the *Origin* from numerous island localities—they have generally flourished "so as to become a nuisance." Although creationism and intelligent design supplied no explanation, such facts were perfectly consistent with colonization by accidental transport. As Darwin noted, amphibians and their eggs are easily killed by immersion in salt water. By contrast, reptiles, which are often found on remote islands, are much better able to survive extraordinary ocean journeys.

The ease with which exotic animals and plants, brought by occasional visitors and colonists, have become naturalized on remote islands provided Darwin with another example of a fundamental weakness in design theory. Almost everywhere that animals and plants have been introduced to ocean islands, the introduced forms have largely exterminated the native forms. When Darwin visited the Galápagos in 1835, this pernicious process had just begun, so he could refer to the looming problem there only in general terms. Commenting on the extreme tameness of the birds and animals in these islands—a disposition that allowed them to be easily killed by the colonists for food or sport—Darwin remarked, "What havoc the introduction of any new beast of prey must cause in a country, before the instincts of the indigenous inhabitants have become adapted to the stranger's craft or power." In support of this Galápagos observation, Darwin drew in the *Origin* on cases of oceanic islands that had been colonized much earlier. Of New Zealand he noted that the endemic organisms were "now rapidly yielding before the advancing legions of plants and animals introduced from Europe," and the original and highly endemic flora of Ascension Island, he reported, was all but extinct.

Had the native animals and plants on those islands been specially designed by God for residing there, surely they ought to have prevailed over introduced organisms, which were presumably designed to live elsewhere. In contrast, evidence showing the consistent superiority of introduced forms was fully in accord, Darwin argued, with the theory of evolution by natural selection, because natural selection acts only on organisms as they compete with one another within local communities. Hence this evolutionary process "will produce perfection, or strength in the battle for life, only according to the standard of that country." Remote islands, which typically have a meager collection of denizens compared with mainland ecological communities, have generally experienced less intense selection over time, and their residents "often yield, as we see they do yield, to the inhabitants of another and generally larger country."

Once Darwin began to catalog instances of imperfect design in nature, more and more examples came to his attention through the revealing lens of evolutionary theory. In the *Origin* he provided scores of additional examples, drawn from biological domains such as morphology, classification, and embryology. The ultimate message in his relentless critique was simple: Any theory of the origin of species had to account not only for adaptations but also for their frequent imperfections. The main problem with the creationist doctrine was the copious evidence of poor design. To account for such evidence, Paley and his predecessors had to assume that their imagined Watchmaker was somehow lacking in foresight and sometimes just plain derelict or even malicious. Darwin, of course, was careful not to push that argument too far, lest he offend his more theologically minded readers. But he did allude more than once in the *Origin* to the dark side of intelligent-design doctrine; for instance, he noted that it seemed preferable to explain the behavior of baby cuckoos ejecting their foster siblings from the nest, or parasitic wasps injecting their eggs into caterpillars (which are then devoured alive by the larvae) as "small consequences of one general law, leading to the advancement of all organic beings" rather than as the outcome of "specially endowed or created instincts." Privately he was more forthcoming. To botanist Joseph Hooker he wrote in 1856, "What a book a Devil's Chaplain might write on the clumsy, wasteful, blundering low & horridly cruel works of nature!"

Ultimately, what Darwin's transformation from creationist to evolutionist reveals about him—and about science generally—is that the best science is conducted in the service of a really good theory. Darwin's own scientific methodology was remarkably modern for a period when Baconian induction—supposedly letting the facts speak for themselves, independently of any theory—was the predominant scientific philosophy. Although Darwin sometimes implied that he was a Baconian scientist in his methods—doubtless to assure people of his efforts to be unbiased—he was anything but a Baconian in practice; throughout his long career he employed what is known as the hypothetico-deductive method, by which hypotheses are used to generate predictions and to guide the collection of relevant evidence—information that is then used either to confirm or reject the hypotheses.[3]

During the *Beagle* voyage, Darwin was guided by creationist theory. As a result of his fateful Galápagos visit and certain other voyage experiences, he discovered that this theory led to false expectations—and also encouraged, as one unfortunate consequence, inappropriate collecting methods. The deeper Darwin probed, the more he realized that creationist theory was abundantly contradicted by the available biological and paleontological evidence. So he rejected this theory and eventually developed a better one, the theory of evolution by natural selection. He then turned this new theory on the same set of phenomena he had once sought to explain by means of creationism and its explanatory handmaiden, intelligent design. Patiently, over the next two decades, Darwin sifted through scientific journals, old tomes about voyages of exploration, gardeners' magazines, and numerous other sources of scientific data and also continually prodded his colleagues for pertinent facts—all in a search for evidence that could be used

---

[3] See especially M. T. Ghiselin, *The Triumph of the Darwinian Method* (Berkeley and Los Angeles: University of California Press, 1969).

to test his theory against the creationist doctrines he had imbibed prior to the *Beagle* voyage.

In 1859 Darwin's revolutionary argument was startling to many, and it provoked great controversy on publication. But it eventually carried the day, as Darwin's scientific contemporaries, following in Darwin's footsteps, addressed the many problematic theoretical issues, highlighted by abundant empirical evidence, that he had raised in the *Origin*. In doing so, most of them realized, just as Darwin had, not only that intelligent design fails to explain anything that cannot be fully explained by natural selection but also—and far more damningly—that whatever this theory does claim to explain, it explains badly or not at all.

In current debates about intelligent design, it is often asserted that this theory is unscientific because it is untestable. The real problem with this doctrine, however, is that numerous straightforward tests have overwhelmingly failed to support it. In lucid and cogent prose, Darwin's "one long argument" in the *Origin of Species* tells us not only why he was personally compelled to reject intelligent design but also why any well-informed person ought to reject it.

FRANK J. SULLOWAY, author of *Born to Rebel: Birth Order, Family Dynamics, and Creative Lives* and *Freud: Biologist of the Mind: Beyond the Psychoanalytic Legend*, is a historian of science and a behavioral scientist at the University of California, Berkeley, where he is a visiting scholar in the Institute for Personality and Social Research and also teaches in the Department of Psychology.

SCOTT ATRAN

Unintelligent Design

SCIENCE IS NOT particularly suited to dealing with problems of human existence that have no enduring logical or factual solution, such as avoiding death, overcoming loneliness, finding love, or ensuring justice. Science cannot tell us what we ought to do or what should be, only what we can do and what is. Religion thrives because it addresses our deepest emotional yearnings and society's foundational moral needs. No society has ever lasted more than a few generations without a moral foundation that, though rationally inscrutable, is considered to be unquestionably true.

In the competition for moral allegiance, secular ideologies are at a disadvantage, for if a better ideology might be available down the line, then there's no better reason to accept the current ideology than convenience. And if people come to believe that all apparent commitment is self-interested convenience—or, worse, self-interested manipulation of others—then commitment withers and dies. In times of vulnerability and stress, the pursuit of self-preservation is likely to take precedence over the pursuit of the public good, as the great Arab historian Ibn Khaldun noted centuries ago. Religion passionately rouses hearts and minds to break out of this viciously rational cycle of self-interest and adopt group interests that can benefit individuals in the long run. In the narrowest case, a couple bound by devotion more easily overcomes personal ups and downs; in the broadest case, a shared faith in an omniscient and omnipotent agent (the supreme being of Abrahamic religions) weakens the every-man-for-himself mentality.

Science, therefore, may never replace religion in the lives of most people, or in any society that hopes to survive for very long. But neither can religion replace science if humankind hopes to unlock nature's material secrets. And parodies of science, like the so-called theory of intelligent design, only cripple science education.

ID advocates assume that a spiritual force, which cannot be reduced to matter and energy, purposefully engineers DNA to produce the variety of living beings and behaviors we see in the world. Notions of ID waffle between a Prime Mover, or architect who stands aside after having given the blueprint for all future developments, and a Divine Creator who actively intervenes in species affairs and human prayers, like a stage director overseeing a play.

Although scientific theories are validated when their surprising predictions are confirmed (as genetics has repeatedly validated nonobvious evolutionary linkages between Earth's creatures), no scientific theory can ever be proved true. Nevertheless, any theory worth its salt can show competing theories to be false—at least over a common set of referents and measures. ID provides no principles or methods to systematically test whether its own assertions are false, much less those of evolutionary theory. But evolutionary theory does provide a logical and factual framework for evaluating the engineering pretensions of ID (whether architectural or directorial), which turn out to be either patently false or so hopelessly vague and open to interpretation as to be scientifically meaningless.

The founders of evolutionary theory saw adaptations—the bullet shapes of fish and sea mammals, the wings of birds and bats, the human being's opposable thumbs and reasoning capacity—as well-crafted designs. Charles Darwin marveled at how adaptations were "highly perfected for any particular habit," and Alfred Russel Wallace saw in them "very much the appearance of design by an intelligent designer on which the well-being and very existence of the organism depends." Still, such designs are far from optimal engineering. This is because there can never be a natural selection of tools and materials from scratch. *Natural selection is always bound by historically antecedent compromises between organic structures and environments.*

On a more abstract level, the causal processes (genetic mutations) that produce new designs in the history of life on Earth are largely random with respect to any functions that those novelties might assume. From an engineering perspective, for example, it's odd that fish species have various numbers of caudal vertebrae, whereas nearly all mammals have just seven cervical vertebrae, the exceptions being the order Xenarthra (sloths, armadillos, and anteaters), with five to nine, and manatees, with six. Odd indeed, if we compare, say, camels and moles.

The ancestral species of the camel family originated in North America millions of years ago, where they evolved a number of adaptations, including a long neck and long legs, to windblown deserts and other nasty environments. Other special designs emerged over time: double rows of protective eyelashes, hairy ear openings, the ability to close the nostrils, a keen sense of sight and smell, humps for storing fat, a protective coat of long and coarse hair (different from the soft undercoat known in the clothing trades as camelhair), and remarkable abilities to take in water (up to 100 liters at a time) and to do without it (up to seventeen days).

Moles, which are an order of magnitude smaller than camels, evolved for burrowing in the earth for earthworms and other food sources that most other animals wouldn't touch. A number of specialized adaptations evolved, but often in directions opposite to those of the camel: round bodies, short legs, a flat pointed head, broad claws on the forefeet for digging, and velvety fur that brushes in any direction without resistance. In addition, most moles are all but blind and lack external ears, which makes burrowing less onerous than if they had functional eyes and external ears.

Given such a stunning variety of specialized differences between the camel and the mole, it's curious that the structure of their necks is basically the same. Surely the camel could do with more vertebrae (swans have twenty-two to twenty-five) and flexibility in foraging through the coarse and thorny plants that are its standard fare, whereas moles could just as surely do with less. What is almost as sure, however, is that there is substantial cost in restructuring the neck's nerve network to conform to a greater or lesser number of vertebrae. The effects of adding or subtracting spinal nerves to conform to addition or subtraction of cervical vertebrae would likely ramify through the upper body (arms, chest, and so forth) in unforeseen and probably dysfunctional ways. For example, one rare variation involving the seventh cervical vertebra is associated with childhood cancer and also with thoracic outlet syndrome, severe pressure on the nerves of the brachial plexus that can lead to the loss of arm use.

Here's another example, closer to home. In land animals, the mouth does double duty, as an opening to take in both food and air. As creatures evolved from water onto land, the opening to the respiratory system was jerry-rigged to share the preexisting digestive tract's anterior structure, including the mouth and pharynx (throat). In terrestrial vertebrates, the pharynx became a short passage linking the mouth to the esophagus and the windpipe. Any mistiming of the swallowing mechanism that blocks off the air passage in routing food to the esophagus causes choking. For humans, the problem is even worse, because the mouth and throat do triple duty, serving also the function of speech. Both in swallowing food and in articulating speech sounds, respiration is temporarily inhibited, as the larynx rises to close (in swallowing) or constrict (in speaking) the opening to the air passage (glottis). Humans are more liable than other animals to choke when they attempt to coordinate eating, breathing, and speaking. In the bargain, the swallowing ability of humans has become much weaker than that of other animals. So when moms tell their kids, "Don't talk while you're eating!" they're helping to make up for an evolutionary failing.

Or consider the procrustean fit of the reproductive, urinary, and excretory tracts to the same anatomical region. In men, the uretha serves both as a urinary canal and a genital duct. This results in dysfunction and diseases that pass from one system to the other. In women, these two passages are anatomically separate but sufficiently close to each other and to the anus to facilitate the spread of infection from each of the three systems to the others, especially during pregnancy.

But the most imperfect design affecting the childbearer's health and life results from evolution's having jammed the outlets of three major expulsive functions into a narrow basin: The expulsion of the large-headed human fetus through a narrow region at childbirth occurs at considerable cost. The "design flaw" of human childbirth has had cascading effects: Human offspring profit from having big brains, but only at substantial cost-to-fitness of relatively high fatality rates for child and mother, long periods of postnatal care, reduction in fertility rates, decrease in resource procurement, and so forth. Many aspects of social life, in turn, may have emerged in natural selection and subsequent cultural selection as compromises to such design problems. Modern societies are still trying to work out these compromises for "the working mother."

Creationists and proponents of ID often point to the human body as evidence of God's plan—of his intelligent design and benevolence toward his creatures. A closer look at our parts reveals that the Deity may never have been wholly pleased with his preferred creations. Why did he invert the retina and give humans (but not the octopus) a blind spot? Why, in making us upright, did he render us so liable to back problems? Why did he give us just one head, heart, and liver, instead of two, like lungs and kidneys? After all, having two lungs and kidneys is surely better than having one of each: If you have one and it fails, you die; if you have two and one fails, you live.

From the Darwinian perspective, any such developments would be catastrophically disruptive for other adapted functions, so that the individuals who had them could not survive. Of course, evolution allows mutations leading to two hearts or heads or livers to occur, but premature death or natural selection weeds them out. Adaptations are constrained by pre-existing structures. Consider the often lethal complications that arise from the birth of a variety of Siamese twins, whose two minds may function prodigiously but whose brains and bodies cannot long survive with only the neural support of a shared spinal cord. Not until very recently have we been able, by means of complicated surgery or genetic engineering, to begin to trump such limitations.

Another limitation on adaptation related to constraint concerns the idea of a "fitness landscape." A functional design in such a landscape represents a "local optimum"—a mountain summit that selection has incrementally reached from a given base. Although neighboring mountains may represent better designs, selection is unable to backtrack down the fitness slope into the deep valleys between the mountains so as to climb to a more advantageous peak. A precipitous descent would reduce fitness dramatically. A gradual descent would amount to successfully reversing the process of natural selection, the probability of which is astronomically low.

Now, two arms and two legs suits human purposes better than the vertebrate base camp of four legs. But why not four arms and two legs, two arms and four legs, or a pair each of wings, arms, and legs? Mythologies the world over intimate the likely functionality of such designs for human purposes. In fact, the insect world provides examples of various designs with three pairs of limbs and one or two pairs of wings. But despite cinematic morphing programs and makeup artistry that can produce the illusion of a human fly or Spider Man, even a small step in that direction would probably fatally compromise an individual's chances of survival. Although genetic engineering has begun to bridge across fitness valleys, some valleys may be too wide for any bridge: Existing surgical techniques could be used to stretch our torso fat and rejigger our ribs to create a wing, allowing us full sensation in our new hanging, boned flaps of flesh. But we couldn't fly with them unless we also had the appropriate neural structure, which would be almost unimaginably complicated either to genetically engineer or to graft into our brains. Only angels, it appears, can have minds wired for wings. Other such designs may be optimal but nearly impossible because there is simply no genetic basis, hence no genetic variation, to start the selection process. That's why there almost certainly can be no Superman with earthly DNA who can hear a whisper miles away, see through solid walls, bore through steel, fly faster than a speeding bullet, or jump tall buildings in a single bound.

Although science may never replace religion, science can help us understand how religions are structured in individual minds (brains) and across societies (cultures) and also, in a strictly material sense, why religious belief (including that in intelligent design) endures. Recent advances in cognitive science, a branch of psychology with underpinnings in evolutionary biology, focus on religion and awareness of the supernatural as a by-product of the convergence of several cognitive and emotional mechanisms that evolved, by natural selection, for the accomplishment of mundane tasks.

But first a caution: Religion *is* a serious riddle for any evolutionary account of human thought and society. Evolution can't explain religion simply as an adaptation for some ancestral task that is "hardwired" into us. There is no gene for the complex of beliefs and behaviors that make up religion, any more than there is a gene for science; nor is there likely any genetic complex with lawlike or systematic qualities that is responsible for most religious belief or behavior. It's rash (and very probably wrong) to infer from the fact that almost every successful human society has been religiously based that nature selected a propensity for religion because ancestral tribes that believed they were favored by the gods were the tribes that beat other tribes (a notion popular in some circles of sociobiology). If you seriously consider what people who believe they are favored by the gods believe in, and how these believers behave as a result, then you've got a good chunk of human belief and behavior to cover.

For example, try to come up with an adaptive logic that could generate a unitary explanation for:

- years of toil to build gigantic structures that house only dead bones (Egyptian, Mesoamerican, and Cambodian pyramids);
- or sacrificing one's sheep (Hebrews) or camels (Bedouin) or cows (Nuer of Sudan) or chickens (Highland Maya) or pigs (Melanesian tribes, Ancient Greeks);
- or dispatching wives when husbands die (Hindus, Inca, Solomon Islanders);
- or slaying one's own healthy and desired offspring (the firstborn of Phoenicia and Carthage, Pawnee and Iroquois maidens, Inca and Postclassic Maya boys and girls, children of South India's tribal Lambadi);
- or chopping off a finger to give to dead warriors or relatives (Dani of New Guinea, Crow and other American Plains Indians);
- or burning your house and all other possessions for a family member drowned, crushed by a tree, or killed by a tiger (Naga tribes of Assam);
- or knocking out one's own teeth (Australian aboriginals);
- or making elaborate but evanescent sand designs (Navajo, northern tribes of Central Australia, Tibetan Buddhists);
- or giving up all sorts of amenities to keep Fridays (Muslims) or Saturdays (Jews) or Sundays (Christians) holy;
- or just stopping whatever one is doing to murmur often incomprehensible words in various postures several times a day.

There are many explanations, in terms of beneficial functions served, for why religion exists. It is said to relieve people's anxieties—but also to increase anxieties, so that elites can exploit them for political purposes. It is said to be liberating and to encourage creativity, and also to stop people from creatively changing or explaining the world. Such functional explanations and their contraries have all been true, depending on the context; yet all also have been true of cultural phenomena other than religion. These monofunctional accounts provide little explanatory insight into psychological selection factors responsible for the ease with which children acquire religious concepts or the ease with which religious practices and beliefs are transmitted from mind to mind. They have little to say about which beliefs and practices are most likely to recur in different cultures and most disposed to cultural variation and elaboration.

Religion has endured in nearly all cultures, and in most individuals *because humans are faced with problems they can't solve*. As people routinely interact, they naturally tend to exploit various mundane cognitive faculties in special ways to solve an array of inescapable existential problems that have no apparent worldly solution. Consider death. Because we have cognitive abilities, we are aware of death everywhere. But physical death is something that organisms have evolved to avoid. So people seek some kind of a long-term solution to this looming prospect while knowing there is none.

Another existential problem is deception, which threatens to undermine the very basis of social trust and group life. Recent experiments show that children as young as three are aware of others having false beliefs. If you have rocks or plants or bodies of water before you and you say, "Oh, there's no rock there," or "That's not wet," or "That's not really a tree," someone can come along and say, "Look, you're nuts; I can touch it; I can show you it's a tree." For ordinary physical events, that is, we have ways of verifying what's real or not. For moral judgments, though, we have nothing comparable. If someone says, "Oh, this one should be a beggar and that one should be a king," or "Murder is bad and capital punishment is good," what is there in the world to prove this is so, or not so? If there's nothing sure, how will people ever get on with one another, especially non-kin? How can they build societies? How can they come to trust one another's loyalty? One solution involves inventing a world governed by deities like big brothers, who watch over the commonweal and make sure that there will be no defectors, no shirkers. As Dostoevski's Ivan Fyodorovitch claims in *The Brothers Karamazov*, since there is "nothing in the whole world to make men love their neighbors," then "for every individual . . . who does not believe in God or immortality, the moral law of nature must immediately be changed into the exact contrary of the former religious law, and . . . egoism, even unto crime, must become not only lawful but even recognized as the inevitable, the most rational, even honorable outcome."

How do people use ordinary abilities to create extraordinary worlds? We create barely counterintuitive worlds that allow supernatural agents to resolve our existential dilemmas—worlds close enough to our factual, everyday world to make them perceptually compelling and conceptually tractable. Religious worlds invariably center on supernatural-agent concepts like gods, goblins, angels, and ancestral spirits. Thus, ghosts and the Abrahamic deity and the devil are pretty much like humans psychologically (belief, desire, promise, inference, decision, emotion) and biologically (sight, hearing, feel, taste, smell, coordination) but lack material substance and its associated physical constraints.

Think about why people around the world are so readily disposed to see gods in the clouds and hear spirits in the wind. In recent years, numbers of devout American Catholics perceived the image of Mother Teresa in a cinnamon bun sold in a Tennessee shop. Latinos in Houston prayed before a vision of the Virgin of Guadalupe, while Anglos saw only the dried ice cream on a pavement. Cuban exiles in Miami spotted the Virgin in windows, curtains, and television afterimages as long as there was hope of keeping young Elian Gonzalez from returning to godless Cuba. And on 9/11, newspapers like the *Philadelphia Daily News* showed an Associated Press photo of smoke billowing from one of the World Trade Center towers that "seems to bring into focus the face of the Evil One, complete with beard and horns and malignant expression, symbolizing to many the hideous nature of the deed that wreaked horror and terror upon an unsuspecting city."

Psychological experiments show that children and adults spontaneously interpret the contingent movements of dots and geometrical forms on a screen as interacting agents with distinct goals and internal motivations for reaching those goals. People will invariably describe a bunch of triangles that converge from all corners of a screen as "pushing" or "herding" a bunch of hapless circles into one corner, when all that's really happening is a change in the distribution of triangles and circles on the screen over time.

Our brains may be wired to spot lurkers and seek protectors where conditions of uncertainty prevail (when startled, at night, in unfamiliar places, during sudden catastrophe, in the face of solitude, illness, death, and so on). Plausibly, the most dangerous and deceptive predator for genus *Homo* since the Late Pleistocene has been *Homo* itself, which may have engaged in an accelerating behavioral and cognitive "arms race" aggravated by individual and group conflicts. Given the constant menace of enemies within and without, the ability to conceal and deceive would favor survival. In potentially dangerous or uncertain circumstances, it would be best to anticipate and fear the worst of all likely possibilities: presence of a deviously intelligent predator. Such a biologically prepared mental faculty (or "cognitive module") would provide a rapid and economical reaction to a wide (but not unlimited) range of stimuli—stimuli that would also have been associated with the actions of furtive predators in ancestral environments. Mistakes ("false positives") would usually carry little cost, whereas a timely response to a real threat could provide the margin of survival. From an evolutionary perspective, it's better to be safe than sorry regarding the detection of agency in mysterious or uncertain circumstances.

Natural selection endows creatures with biological and mental faculties that allow them to respond adequately but by no means perfectly to their environment. Consider food-catching behavior in frogs. When a flying insect moves across the frog's field of vision, bug-detector cells are activated in the frog's brain; these cells in turn massively fire others, in a chain reaction that usually results in the frog's shooting out its tongue to catch the insect. The frog's evolved bug-detector is primed to respond to any small dark object that suddenly enters the visual field, so you can trick a frog into slurping up little black wads of paper dangling before it just as if they were flies, even to the point of killing the frog. Humans take advantage of this sort of imperfection in evolution by habitually "fooling" themselves, as when people become sexually aroused by makeup (which artificially highlights sexually appealing attributes), fabricated perfumes, or undulating lines on paper or dots arranged on a computer screen (that is, pornographic pictures). Indeed, much of human culture (for better or worse) can arguably be attributed to focused stimulations and manipulations of our species' innate proclivities. Such manipulations can serve cultural ends far removed from the ancestral adaptive tasks that gave rise to the cognitive and emotional faculties triggered (as in the sultry advertisements for deodorants, dishwashing powder, or a French Arabica "coffee named desire").

But why do people accept supernatural worlds that contradict everyday facts and logic and demand personal sacrifice? Because this collectively satisfies the emotions that motivate religion in the first place. Communal rituals coordinate emotional validation of and commitment to moral truths, in worlds governed by supernatural agents. Pageantry and religious ritual coordinate their participants' minds and bodies into convergent expressions of public sentiment—a sort of bonding (not unlike mutual courtship displays in the animal kingdom, or a mother rocking her child) that turns moral consensus into something sacred, transcending all reason and doubt. For those left outside the consensus, though, cruelty and intolerance are often all that religion has to offer.

The rituals accompanying religions almost always include music and other sorts of voluntary rhythmic stimulations. Even the Taliban, who banned nearly every sort of collective sensory stimulation, systematically used *a capella* chants to cement adherence to their religious fraternity. Just about all religions employ submissive gestures in prayer: outstreched arms with chest exposed and throat bared, genuflection, prostration, and so on. These are pretty much the same bodily expressions other social animals use to signal submission, including our simian cousins and canine friends. The emotional unity created during these performances underscores the commitment to sacrifice something of oneself for others—not necessarily in the here and now but as an open-ended promise to help others when they may need it most. From this commitment, humanity has benefited enormously.

To evoke a strong and intimate sense of community, members of religious sects often create families of fictive kin. For example, the mujahedin I have interviewed—from Paris suburbs to Indonesian jungles—who profess commitment to martyrdom come from almost every walk of life and socioeconomic background (which is why global profiling across cells of suicide bombers is a waste of time); however, within each cell of typically eight to twelve people, all tend to eat the same sort of food, wear the same sort of clothing, chant the same slogans, and share the same daily rites and routines. Through incorporation of recruits into relatively small and closeted cells—emotionally tight-knit brotherhoods—religiously inspired terror organizations create a family of cellmates who are just as willing to sacrifice for one another as a parent is for a child. These culturally contrived loyalties mimic and (at least temporarily) override genetically based fidelities to kin, while securing belief in sacrifice for a larger group cause. This solidarity can even mutate, over the Internet, into a virtual family, whose members may be physically remote from one another but who are as emotionally linked and primed for mutual sacrifice as any real family or group of friends.

Faith in intelligent design is a milder convergence of these sorts of cognitive and emotional proclivities onto a plane of existential anxiety. Advocates of ID, including the Pope and the Dalai Lama, consider the evolutionary thesis that life evolved by random combination of genes (along with nonrandom cumulative natural selection of the fittest) to imply arbitrariness in moral choice and thus to invite nihilism and spiritual poverty. But this anxiety is unwarranted: Nothing indicates that people who believe that life arose by chance also believe that morality is haphazard.

As Americans rose against England, inspired by Benjamin Franklin's credo that "rebellion against tyrants is obedience to God," the English historian Edward Gibbon began writing

Decline and Fall of the Roman Empire, attributing Rome's collapse to religious infection by Christianity. Ever since, most politicians (and ordinary people) have continued to praise God, whereas most scientists and secular-minded scholars have continued to bemoan religion's influence and predict its demise. Religious fervor is, if anything, increasing across the world, including in the United States, the world's most economically powerful and scientifically advanced society. An underlying reason is that science treats humans and their intentions only as incidental elements in the universe, whereas in religion (as for people generally) they are central. That's why Thomas Jefferson's impersonal unitarian God and the French Revolution's neutral deity fell by the wayside, and why as many as half the people in the Soviet Union professed religious belief. Personal gods speak to personal problems—perhaps even more so in complex and mobile societies increasingly divorced from nurturing family settings and long-familiar environments.

But in purging intentional causes, including supernatural agents, from science, great progress has been made not only in understanding nature but in ameliorating the human condition, as with modern medicine. Those who preach intelligent design would reintroduce intentional cause into science and so reduce science's ability to serve. For society, that is an unintelligent design.

SCOTT ATRAN is directeur de recherche (Anthropology), Centre National de la Recherche Scientifique in Paris and adjunct professor of psychology at the University of Michigan.

STEVEN PINKER

Evolution and Ethics

I SUSPECT THAT most people who have reservations about the teaching of evolution don't really care about the best theory of where zebras and dandelions come from. They care about meaning and morality. They believe that the theory of evolution corrodes the foundations of ethics, which they take to be the acceptance of God's purpose, the following of God's laws, and the anticipation of divine retribution in an afterlife. These beliefs, they argue, are all that stands between us and a life of amoral callousness and hedonistic nihilism. As the slogan goes: "If you teach children that they are animals, they will behave like animals."

Many proponents of "intelligent design" have made it clear that they feel that their campaign is ultimately about the basis of morality. Even some scientists who are militantly opposed to the teaching of creationism in schools, such as the late Stephen Jay Gould, have argued that religion is the only legitimate source of ethics—that science and religion are "non-overlapping magisteria," with science having authority over matters of fact and religion having authority over matters of meaning and morality.

The point of this essay is that this line of thinking is mistaken. It's true that science cannot provide us with moral principles. But neither can religion. An understanding of morality is to be found through secular moral reasoning and lies in fundamental facts about the human condition, not in the dictates of a supernatural deity.

It's surprising that so many people take seriously an association between religion and morality in the first place. A glance at history and current affairs suggests the opposite. The Bible contains several injunctions from God to the Israelites to slay the occupants of towns they covet—except for the young women, whom they are to take as unwilling wives. Since then, religions have given the world stonings, witch burnings, crusades, inquisitions, holy wars, jihads, fatwas, suicide bombers, gay bashers, abortion-clinic gunmen, child molesters, and mothers who drown their sons so they can happily be reunited in Heaven.

Defenders of religion as a source of morality say that these crimes are exceeded by the atrocities of atheistic ideologies, like Nazism and Marxism. This is based on an error and a non sequitur. The error is the belief that Nazism was based on atheism; though Hitler was ambivalent about Christianity, he was not an atheist and may have been emboldened by the conviction that he was carrying out a divinely ordained plan. The non sequitur is the belief that there are two possible sources of morality: religion and atheism. If atheism isn't enough to make people moral, religion must fulfill that role. It is a non sequitur because atheism is defined by negation—an unwillingness to swallow certain beliefs about the supernatural—and has no content on its own. Yes, atheism cannot prevent mass murder or other forms of evil, for the same reason that it can't build bridges, cure cancer, or put a human being on the moon. Why would anyone think it could? The question is not what atheism by itself can accomplish (which is nothing, except to eliminate false starts). The question is, What *can* accomplish what we want? The fact that atheism can't build a bridge or cure cancer doesn't mean that religion can do those things; it means that we have to look to engineering and science. And the fact that atheism can't guarantee that people will be moral doesn't mean that religion can, either; it means that we have to look to moral values and moral reasoning.

How can one be so sure that morality is separate from religion? The argument is straightforward and goes back to Plato. Assume, for the sake of argument, that moral values come from God's commandments. One can then ask the question, Why did God choose those commandments, as opposed to other ways he could have commanded us to behave? There are two possible answers. One is that he had no reason; the commandments are arbitrary whims. In that case, one should ask, Why should we follow them? The other answer is that God did have a good reason. In that case, why shouldn't we appeal to the reasons directly? What does God's signing onto them add?

The point can be put another way. What should you do if God commands you to carry out an act that is patently evil, such as tying up a child and stabbing it to death? If the source of morality is religion, you should go ahead and do it. If, in contrast, you have good reasons for *not* doing it, that means that God's commandments are not the source of morality. It won't do any good to say that God would never issue such a commandment. He did so with Abraham, and Abraham's willingness to carry it out is, bizarrely, often held up as a virtue. From the genocides in Canaan sanctioned in the Bible to the Islamist suicide bombers today, we find numerous examples of God's commanding people to do patently evil acts.

The recurrence of atrocities committed in the name of God shows that they are not random perversions. Since unverifiable beliefs have to be passed along from parents and peers rather than discovered in a world we all share, they differ from group to group and become divisive identity badges. And an omnipotent authority that no one can see is a useful backer for malevolent leaders hoping to enlist holy warriors.

For the same reason, we must reject the idea that a fear of divine retribution in an afterlife is necessary to deter people from committing evil acts when no one is looking. Perhaps it is true that if nonbelievers thought they could elude the legal system, the opprobrium of their communities, and their own consciences, they would not be deterred by the threat of spending eternity in hell. But they would also not be tempted to blow up thousands of people by the promise of spending eternity in Heaven. The basic problem is that our own eyes tell us that virtue is not rewarded, nor evil punished, by miraculous acts in this world. So we have to rely on proclamations by religious authorities about how divine justice will be meted out in the next world. And that puts a mighty temptation in their path.

If morality doesn't come from religion, where does it come from? This is really two questions. One of them is, Where does our conscience or moral sense come from? The other is, Do moral principles come only from our moral sense, and therefore might they be mere figments of our neurobiological makeup, or can they be considered an objective part of reality?

That *Homo sapiens* has a moral sense is undeniable. Children as young as a year and a half spontaneously give toys, proffer help, and try to comfort adults or other children who are visibly distressed. People in all cultures distinguish right from wrong, have a sense of fairness, help one another, impose rights and obligations, believe that wrongs should be redressed, and proscribe rape, murder, and some kinds of violence.

Contrary to the cutthroat connotations of the word "Darwinian," the theory of natural selection has no trouble explaining the evolution of a moral sense. The love of one's children and other blood relatives is a direct consequence of the way evolution works. A gene that impelled an organism to help a close relative would indirectly be helping a copy of that gene sitting inside the relative, and such a gene would then spread through the population. This is sometimes explained by the metaphor of genes being "selfish" agents that "try" to get copies of themselves into the next generation, but it's crucial to remember that it's just a metaphor. The genes can be said to have metaphorical motives—making copies of themselves—and the organisms they design have real motives. But they are not the same motives. Sometimes the most selfish thing a gene can do is wire *unselfish* motives into a human brain—heartfelt, unstinting, deep-in-the-marrow unselfishness. As I wrote in my 1997 book *How the Mind Works*:

> The body is the ultimate barrier to empathy. Your toothache simply does not hurt me the way it hurts you. But genes are not imprisoned in bodies; the same gene lives in the bodies of many family members at once. The dispersed copies of a gene call to one another by endowing bodies with emotions. Love, compassion, and empathy are invisible fibers that connect genes in different bodies. They are the closest we will ever come to feeling someone else's toothache. When a parent wishes she could take the place of a child about to undergo surgery, it is not the species or the group or her body that wants her to have that most unselfish emotion; it is her selfish genes.

Natural selection can also install selflessness in couples. If a couple were guaranteed to be faithful, to favor each other over their kin, and to die at the same time, their genetic interests would be identical, wrapped up in the single basket containing their common children. Under that idealization, the love between a man and a woman should be perfect, each willing to undergo any sacrifice for the other. Of course the idealization is nowhere near true and, correspondingly, marital love is not perfect. But its existence in the first place is neither in doubt nor evolutionarily inexplicable.

What about people who are not related by blood or marriage? Biologists have also long understood the evolution of moral sentiments toward non-relatives. One organism helps another by grooming, feeding, protecting, or backing him, and is helped in turn when the needs reverse. This is called reciprocal altruism, and it can evolve because cooperators do better than hermits or misanthropes. Reciprocators enjoy the gains of trading their surpluses, saving each other from drowning or starvation, babysitting each other's children, and enjoying countless other opportunities to help or avoid hurting. They can also do better over the long run than the cheaters, who take favors without returning them—as long as the reciprocators come to recognize the cheaters and shun or punish them.

The result is a suite of moral emotions. Sympathy and trust prompt people to extend the first favor. Gratitude and loyalty prompt them to repay favors. Guilt and shame deter them from hurting or failing to repay others. Anger and contempt prompt them to avoid or punish cheaters. And among humans, any tendency of an individual to reciprocate or cheat does not have to be witnessed firsthand but can be recounted by language. This leads to an interest in the moral reputation of others, transmitted by gossip and public approval or condemnation, and a concern with the moral esteem in which one is held by others.

Of course, people do more for their fellows than return favors and punish cheaters. They often perform generous acts without the slightest hope for payback, ranging from leaving a tip in a restaurant they will never visit again to throwing themselves on a live grenade to save their brothers in arms. But as several biologists and economists have pointed out, some degree of pure magnanimity can evolve as well. What it takes is that other people seek to discriminate fair-weather friends from loyal allies. Signs of heartfelt loyalty and generosity serve as guarantors of one's promises, reducing a partner's worry that you will default on them. The best way to convince a skeptic that you are trustworthy and generous is to *be* trustworthy and generous.

So not only is Darwinism compatible with our having a moral sense; it can explain why we evolved to have one. It can also explain the limitations of our moral sense and its ability to develop over time. Any understanding of human morality has to explain the moral progress that has taken place over the millennia. Customs that were common throughout history and prehistory—slavery, punishment by mutilation, execution by torture, genocide for convenience, endless blood feuds, the summary killing of strangers, rape as the spoils of war, infanticide as a form of birth control, and the legal ownership of women—have vanished from large parts of the world.

Continuous moral progress is not incompatible with the existence of a fixed moral sense. Moral improvement can be understood in terms of what the philosopher Peter Singer calls "the expanding circle." People have steadily expanded the mental dotted line that embraces the entities considered worthy of moral consideration. The circle has been poked outward from the family and village to the clan, the tribe, the nation, the race, and most recently (as in the Universal Declaration of Human Rights) to all of humanity. This expansion has happened for a number of reasons. As people in more parts of the planet become economically interdependent, the hatred between them decreases, for the simple reason that you can't kill someone and trade with him too. Our sympathy response can also be cranked up by new kinds of information demonstrating that other folks are similar to ourselves. Words and images from erstwhile enemies can make it impossible to continue to believe that they are subhuman. A historical record can warn against self-defeating cycles of vendetta. A cosmopolitan awareness, enriched by travel, education, and exposure to realistic fiction may lead people to think, "There but for fortune go I." An expansion of sympathy may come from something as basic as the requirement to be logically consistent (and hence taken seriously) when imploring other people to behave in certain ways. People come to realize that they cannot force others to abide by rules that they themselves flout. Egoistic, sexist, racist, and xenophobic attitudes are logically inconsistent with the demand that everyone respect a single code of behavior.

This brings me to the second question about morality. Is morality a mere hallucination of our moral sense, like the fact that we like the smell of roses and dislike the smell of carrion (while flies, presumably, have the opposite reaction)? In fact, a biological understanding of morality is consistent with the possibility that morality is more than this—that morality has some external warrant in the nature of reality.

Many of our faculties evolved to mesh with real entities in the world. Our perception of depth is the product of complicated circuitry in the brain, but that does not mean that there aren't real rocks and cliffs out there. And so it may be with certain abstract entities. Humans, like many animals, appear to have an innate sense of number, which can be explained by the advantages of reasoning about numerosity during our evolutionary history. (For example, if three bears go into a cave and two come out, is it safe to enter?) But the mere fact that a number faculty evolved does not mean that numbers are products of the mind. According to the Platonist conception of number favored by many mathematicians and philosophers, numbers and shapes have an existence independent of minds. The number 3 is not invented out of whole cloth; it has real properties that can be discovered and explored. No rational creature equipped with circuitry to understand the concept "2" and the concept of addition could discover that 2 plus 1 equals anything other than 3. That is why we expect similar bodies of mathematical results to emerge from different cultures or even on different planets.

A similar argument can be made for morality. According to the theory of moral realism, right and wrong exist and have an inherent logic that licenses some moral arguments and not others. The world presents us with non-zero-sum games in which it is better for both parties to act unselfishly than for both to act selfishly. True, I might be best off if I acted selfishly and everyone else acted unselfishly (at least toward me), but unless I am Galactic Overlord who rules over everyone in the universe, I'm unlikely to get that wish. I am better off if no one hurts me, but I can hardly demand that others refrain from doing so if I am not willing to forgo hurting others. And since I am better off not hurting and not getting hurt than hurting and getting hurt, I am compelled to insist on a moral code, even if the price is adhering to it myself. An amoral, thoroughly self-interested philosophy falls apart as soon as one sees oneself from an objective standpoint as a person just like others. It is as illogical as insisting that "here," the point in space one happens to be occupying at the moment, is a special place in the universe, just because one is standing on it.

So our moral sense may have evolved to mesh with an intrinsic logic of ethics rather than concocting it in our heads out of nothing. It is in the very nature of moral discourse that arguments are forced to go in certain directions as soon as one tries to prevail upon others as to how they "should" behave. No creature equipped with circuitry to understand that it is immoral for you to hurt me could discover anything but that it is immoral for me to hurt you. As with numbers and the number sense, we expect moral systems to evolve toward similar conclusions in different cultures or even on different planets. That is why the Golden Rule has been rediscovered many times—not just in religious traditions in several parts of the world but in numerous secular arguments such as the social-contract theories of Hobbes, Rousseau, and Locke; by Kant in his categorical imperative; by Singer in his expanding circle; and by John Rawls in his notion that we should agree to a distribution of social positions as if we were under a "veil of ignorance" about our personal traits. They are all versions of the same basic principle: the interchangeability of perspectives—or "view from nowhere," as the philosopher Thomas Nagel has put it—which is the real basis for morality.

In sum: Neither scientists nor defenders of religion should accept the premise that morality belongs in the province of religion. In practice, religions have sanctioned vile acts throughout human history. In principle, there is no reason to believe that religious edicts are inherently moral or that they foster morality, and some reasons to believe that they can have the opposite effect. By contrast, an evolutionary understanding of the human condition, far from being incompatible with a moral sense, can explain why we have one. Natural selection can favor a moral consideration of relatives and trading partners not just as a cynical strategy for self-betterment but as a heartfelt conviction. Moral progress over the millennia is explicable in terms of the circumstances that encourage people to project themselves into the lives of others and hence expand the circle of empathy. And morality is not necessarily a mere secretion of an evolved moral sense but may have an inherent logic (rooted in the interchangeability of vantage points) that the moral sense evolved to mesh with.

None of this should be seen as a blanket condemnation of religion. Religions have provided comfort, community, and wisdom to countless people throughout history and have contributed enormous bodies of art and scholarship to humanity. But people should resist the idea that religious reservations about the teaching of evolution have any moral weight in their favor, or that a modern understanding of evolution by natural selection erodes the foundations of morality.

STEVEN PINKER, a cognitive scientist and the Johnstone Family Professor in the Department of Psychology at Harvard University, conducts research on language and cognition. He is the author of numerous books, including *The Language Instinct, How the Mind Works, Words and Rules,* and *The Blank Slate*.

LEE SMOLIN

Darwinism All the Way Down

DARWINIAN EVOLUTION by natural selection is certainly settled science in biology. But this does not mean that we are done absorbing the implications of Darwinian natural selection for science. Consider, for example, the following reflection, written in 1891 by Charles Sanders Peirce, one of the inventors of the American pragmatist school of philosophy:

> To suppose universal laws of nature capable of being apprehended by the mind and yet having no reason for their special forms, but standing inexplicable and irrational, is hardly a justifiable position. Uniformities are precisely the sort of facts that need to be accounted for. Law is *par excellence* the thing that wants a reason. Now the only possible way of accounting for the laws of nature, and for uniformity in general, is to suppose them results of evolution.

When I went to university in the 1970s, I was told that the goal of physics was to discover what the laws of nature are. At about the same time, a theory arose that has functioned ever since as a basic law of nature, called the standard model of particle physics. This theory seems so far to encompass correctly all observed phenomena save for gravitation and cosmology. It has held up so well that during the intervening three decades we have been bold enough to ask a different question: Why these laws? Why does the universe obey the particular laws we observe it to, rather than different laws?

This is the question Peirce was asking back in 1891. As he certainly knew, it is easy to give answers to this question that point outside of science. The most common example is to attribute the choice of the laws to a God who made the universe. But Peirce was asking for a scientific answer to this question. A scientific answer is one whose truth can be investigated by deriving predictions for experiments from it and then doing the experiments to see if theory is confirmed or falsified.

What Peirce is saying is that the only possible scientific answer to the question of "Why these laws?" is one that explains the laws as having arisen by a process of Darwinian evolution. If this is true, then the ultimate significance of Darwinian evolution goes beyond biology—to the possibility of moving science to a new stage, in which we investigate not only *what* the laws of nature are but *why* they are.

The idea that laws may evolve is a huge departure from previous expectations. At least since Newton, we have gotten used to conceiving of the laws of nature as eternally true. They are supposed to exist in a kind of timeless, Platonic realm along with other timeless truths, such as those of mathematics. This naive view is currently challenged on several fronts. The most direct challenge comes from cosmology itself, where we are forced to wonder what it can mean for a law to be *eternally* true if the universe to which it applies is less than 15 billion years old. Another blow to the Platonic view of laws is provided by general relativity. It used to be thought that one of the laws of physics was that the geometry of space is the simple Euclidean geometry we learn in school. But general relativity tells us that the geometry of space is contingent and dynamical. So an aspect of the world we used to think of as law turns out to be a contingent fact that is the result of evolution in time. This leads us to wonder what else of what we are used to thinking of as eternal law is actually the contingent result of dynamical evolution.

But this kind of unease would lead nowhere, were it not for Darwin's great realization that order can arise from a process of evolution. This gives us a new possibility, which is to consider that the order captured by laws of nature is the result of a process of self-organization acting at the most fundamental scales.

If this is true, then Darwin's revolution does more than push the role of the creator to one side: It undermines the whole philosophical and conceptual framework in which Western thought has operated, within which the more true a statement is, the more it is supposed to be about a transcendent and timeless realm rather than the contingent and fleeting reality we experience. Darwin's great discovery undermines this, because it offers us a rational and testable explanation for how the laws of nature may have arisen, which then requires us to see the laws as mutable and time-bound.

To understand the import of this idea, let's consider that many of our discussions about the relationship between Darwinian evolution and religion are asking the wrong question. It should not, after all, be surprising if people who believe that all truth comes from an ancient text disagree with Darwin, whose ideas are in no ancient text. Rather than bemoaning the fact that fundamentalists disagree with Darwin, let's ask a much more interesting and disturbing question: Why do so many nonfundamentalist theologians and religious leaders have no trouble incorporating Darwin into their worldview?

The reason is that when Darwin's insight is applied only to biology, there remains a big role for an intelligent and benevolent creator. There is after all no need for an omniscient God to have troubled himself intervening in the details of the evolution of life. Being infinitely intelligent, he would have anticipated the power of natural selection and realized that to get a universe with intelligent life, all he had to do was set up the laws and the initial conditions of the universe correctly. Once this was done right, life was bound to originate and evolve. Belief in Darwinian biology and belief in an intelligent creator remain perfectly compatible.

Indeed, many would go further and claim that there is *evidence* for a God in the laws of physics. Suppose one finds evidence for design in the laws of nature by showing that they are in some sense highly improbable. It would be hard to argue that improbable design is the result of a natural process. Natural processes take time, and laws are eternally true, so they must have been true as far back in the past as the past goes. Thus the argument from design may be posed more successfully than it has been for biology by applying it to the universe as a whole.

In fact, in recent decades, just this kind of argument for an Ultimate Designer has been put forward by some very admired cosmologists and astrophysicists. It goes like this: The standard model of particle physics has some twenty free parameters (such as the masses of the quarks, the electron, and the neutrino), whose values may be freely specified. The actual values we assign to them were determined by experiment, but the theory appears to make no demands on the choice and to be equally consistent if other values are chosen. Hence the standard model describes an infinitude of possible universes, only one of which is our own.

Now, one can ask how many of those other universes resemble ours. The answer is very few. Most of the universes described by different choices of the parameters appear boring compared with ours, typically having no stable nuclei other than hydrogen and no long-lived, stable stars. As far as we can tell, they would contain no galaxies, no stars, no chemistry, and certainly no life. Many would consist of nothing but hydrogen in thermal equilibrium; others would just be a gas of neutrons.

It turns out that the parameters we observe have an extremely rare property, which is that they lead to a universe with around 100 kinds of stable nuclei. This makes possible both complex chemistry and long-lived stars, two necessary conditions for life. The parameters that make a universe hospitable to life appear to inhabit a tiny range.

It turns out that this elementary-particle story is repeated in cosmology. There are roughly fifteen parameters specifying the age, size, and structure of our universe. Were these not tuned to an improbably narrow range of values, there would be no galaxies, stars, or life.

The discovery of *the improbable biofriendliness of the universe and its laws* leads to two urgent questions:

- Why do we find ourselves in a universe with the improbable values of parameters necessary for our existence?
- How did it come to be that these particular parameters were chosen?

The first question has an obvious answer: There is no other kind of universe we could find ourselves in. But this obvious fact does not give us an answer to the second question. The second question asks for a mechanism to explain why a universe that satisfies these laws exists at all. This is the question Peirce was addressing. Ignored for more than a century, it has recently become one of the most urgent questions in physics and cosmology. The answers so far proposed for it fall into four groups.

The first answer is the theistic one. It asserts that there is a benevolent God who made the universe, and he made it as a home for life. In modern usage, this is called the *strong anthropic principle*. This principle offers a role for an intelligent designer. One may even argue, with the distinguished South African cosmologist George Ellis, that God set up the whole game, including the choice of laws that are equally consistent within a large range of possible parameter values. In such a world, the intelligent beings that natural selection would eventually produce would come to understand the incredibly improbable circumstances that made their existence possible and would arrive by rational argument at the discovery that the world must have been made by a benevolent creator.

Of course, this line of argument can succeed only if there are no alternative explanations for the biofriendliness of the laws. But there are alternatives.

The second answer claims that there is only one mathematically consistent possibility for the ultimate laws of nature, and it leads to the choices of the parameters we observe. This may be called the *math-tropic principle*. Were this realized, we would all end up as mystics, worshiping some beautiful mathematical fact as the reason for our existence. But for better or worse, this is an alternative that has been tried and on present evidence has failed. The most complete proposal yet made for a consistent unification of all the forces of nature is string theory. There is evidence that it comes in an infinite number of versions, most of which have many free parameters.

The third and fourth explanations share a common and radical feature: They posit that our universe is one of a very large population of universes. We call them *multiverse* theories. They share another assumption, which is that the choices of parameters of the laws of physics vary among the universes in the multiverse. But aside from this, they differ profoundly.

In the third kind of answer, the distribution of parameters over the universes is random and static. This is typically because the universes have no relation to one another. In these scenarios, all the universes are born from a single primordial state. That state exists forever, and the proportion of different kinds of universes in the ensemble is static and eternal. An example of such a theory is eternal inflation: In it, a vast population of universes is being continually born—with an unchanging distribution of parameters—from an eternal and unchanging vacuum state.

Since the choices of parameters that make a universe biofriendly are highly improbable, it follows that our universe is a highly atypical member of this population. Most members of the multiverse have no galaxies, stars, or chemistry, but if there are enough universes, then even a very improbable one like ours will exist. Once we accept the assumptions of this scenario, we have an explanation for the existence of a biofriendly universe. This kind of explanation is sometimes called the *weak anthropic principle*.

But here is the rub: If the choice of parameters is random, then we can get no better understanding of why there are biofriendly universes other than that there is room in the population for every possible universe. There can be no further explanation for the features of the universe we inhabit than that it's one of the biofriendly ones—because it *must* be, in order for us to live here. Such theories necessarily leave much we would like to know about our own universe unexplained. Suppose we wish to understand the value of some parameter *X*, whose value does not alter biofriendliness. Since the constants are distributed randomly in the multiverse, there will be biofriendly universes with every possible choice of *X*. So we can neither predict nor explain the value that *X* will take in our universe. The only exceptions to this lack of predictability will be parameters whose value must be fixed in order for the universe to be biofriendly. So the hypothesis that there is a vast population of universes has no predictive power, because any prediction the theory makes must be a consequence only of the necessary fact that life can exist only in a biofriendly universe. Hence, the hypothesis that this theory is based on—that there exists a vast ensemble of universes with randomly distributed parameters—is both unconfirmable and unfalsifiable. Thus it cannot be considered a scientific theory.

Some physicists argue that even if this hypothesis cannot be tested scientifically, it might still happen to be true. Indeed, one cannot prove that our universe is not one of a vast population of universes among which the parameters are distributed randomly. But many things might be true that cannot be proved or disproved. The main lesson of science is that we make progress by considering only hypotheses that can be either proved or disproved by rational argument from the evidence.

The weak anthropic principle does offer one philosophically important feature as an alternative to the strong anthropic principle. Because it is logically possible, it gets us out of any argument that would require us to believe in a benevolent creator. But this is an illusionary victory for science, because, as we have seen, the weak anthropic principle shares with the strong anthropic principle the unfortunate feature that its hypotheses are neither confirmable nor falsifiable. Indeed, for the purpose of making predictions, both versions of the anthropic principle have the same logical form. Both posit something much vaster than our universe as a means to explain its biofriendliness. In each case, there is no experiment or observation that could either confirm or falsify the hypothesis, because it is consistent with any possible outcome of any future experiment. Whether that untestable hypothesis is the existence of a benevolent creator or a vast population of universes with random laws makes no logical difference to the argument.

Luckily there is a fourth way to answer the question of how the universe came to be biofriendly, and it is a genuine scientific theory. This is to construct a multiverse theory within which our universe is a typical member of a highly non-random population, rather than an improbable member of a random population.

How are we to do this? The only way I know is to follow the advice of Peirce and extend the logic of Darwinism to the universe and the laws of physics. To apply the logic of natural selection to cosmology, one needs the following elements:

(1) A process by which universes reproduce themselves;
(2) A process by which the parameters of the standard model of particle physics change by small, random increments in each new universe produced. These first two elements result in a population of universes with a distribution of parameters.
(3) A mechanism of selective pressure that will result in a highly non-random distribution of parame-

> ters. This happens if the number of progeny a universe has is strongly dependent on the choices of parameters.

Unlike the hypotheses involving the weak and strong anthropic principles, this fourth hypothesis can result in a theory falsifiable by observations—because the population of universes created by a process of natural selection will be far from random. Our universe will then be seen as a *typical* member of this very untypical collection of universes. What we need to determine, in order to test the theory, is only that almost every other universe in the collection has a property not required for our existence.

This strategy provides a possible answer to the question of why our universe is biofriendly that is falsifiable by observation. Here we will rely on one of the oldest hypotheses about quantum spacetime—a hypothesis arising from the combination of quantum theory with Einstein's theory of general relativity. Given very general and plausible assumptions, it turns out to be inevitable that there are collections of events in the universe where all physical quantities become infinite—all at once—all over a region of space. These are called singularities. One such singularity occurs in the very early universe. This is a singularity that lies in our past. There are also singularities in our future. There is at least one at the center of every black hole in the universe.

In the 1960s, John Archibald Wheeler and Bryce DeWitt, two pioneers of the field of quantum gravity, hypothesized that quantum effects would remove these singularities—and that therefore time extends back before the Big Bang. Indeed, it can be argued that before the Big Bang the universe was larger and was contracting toward a singularity. Quantum effects removed that singularity, causing the universe to begin expanding again. We call this a bounce. The same thing was conjectured to happen to the singularities in black holes. They would bounce, and time would continue into the future, into a new expanding universe. Thus, it is natural to hypothesize that the Big Bang in our past was the result of a black hole in another universe.

There has recently been a lot of progress in the field of quantum gravity; in the last five years, we have been able to attack the problem of what happens to singularities. The calculations strongly support the hypothesis of DeWitt and Wheeler that singularities are replaced by bounces.

Our universe has at least 10^{18} black holes, formed by remnants left over by the stellar explosions known as supernovas. Thus we have good theoretical evidence that there is indeed a multiverse consisting of a population of universes that are related to one another genetically—that is, each universe, including our own, has ancestors and progeny.

The evidence I mentioned is, however, theoretical. Could this theory make a prediction that can be tested? Let's follow the story a bit further.

Wheeler proposed that when a new universe was made, the parameters of the laws of physics were changed. He called this "reprocessing the universe." To have a scenario in which natural selection operates, we need only add the condition that, on average, in each generation these changes are small.

There is some theoretical evidence that this is the case. It appears that the values of many of the parameters are dependent on the solution of the theory of elementary particles, and can vary in time. The events in which the parameters vary are phase transitions, like the changes from ice to water to gas. The cosmological situation is closely analogous to that simple example, except that there may be a large number of possible phases, within which the parameters differ. A transition between these phases requires a lot of energy and so will not happen in normal circumstances. But the violent bounce between universes is exactly right to catalyze such a change of phase.

Thus we have both conditions (1) and (2). We get from them a population of universes with a distribution of parameters. Do we have the third condition as well?

We do, because it is not hard to show that the number of black holes depends strongly on the values of the parameters. Most of the black holes in our universe are the relics of very massive stars—stars at least twenty times heavier than our sun. Such stars are rarer than our sun, but not all that rare. If one looks into how such stars get made, one finds a wonderful surprise: The processes that astronomers believe necessary to assemble that much mass require the existence of carbon and oxygen. This means that a biofriendly universe is also a black-hole-friendly universe! Universes that aren't biofriendly won't have black holes, and therefore the biofriendly ones predominate.

Thus this theory provides a genuine explanation for the biofriendliness of our universe—an explanation that is not a tautology. But can we tell whether or not the theory is true? To test it, we must use it to make a prediction about some feature of our universe that has nothing to do with the existence of life—that is, there must be some feature that increases the numbers of black holes without affecting the existence of life. If there are two possible universes, both of which have life but one of which has many more black holes than the other, then the theory predicts that we should find ourselves in the one with the greater number of black holes. This is because the more a universe reproduces, the more universes like it are represented in the multiverse.

Here is an example of a prediction made using this logic. It rests on the fact that remnants of supernovas have two possible fates: If they are light enough, they become neutron stars; if they are sufficiently massive, they become black holes. There is a threshold between these two outcomes. To have as many black holes as possible, the threshold should be as low as it possibly can be without affecting the processes that make the carbon and oxygen that both we and the massive stars require. It turns out that there is a parameter that does control the height of the threshold, without affecting biofriendliness. This is the mass of the strange quark. I will not go here into how it works, but the result is a clean prediction: The threshold must be no higher than 1.7 times the mass of the sun—that is, no neutron stars can be any heavier than that.

So far, this prediction is holding up. All well-measured neutron-star masses are between 1.3 and 1.45 times the mass of the sun. But a heavy neutron star could be discovered anytime. If a neutron star is found with a mass of, say, twice that of the sun, this theory will be ruled out—because it would then be possible to make a universe as biofriendly as ours but with many more black holes.

Thus, we have found a theory that both explains why our universe is biofriendly and is falsifiable. We have done it by following Peirce's advice and hypothesizing a scenario within which the laws of nature are the result of evolution by natural selection.

We can draw some conclusions just from the existence of this theory.

First, we see that a creator is not necessary to explain the biofriendliness of our universe. Either the second or the fourth scenario provides an explanation for the existence of life that has no place for an intelligent designer.

Second, for science, there are clear advantages of the fourth Darwinian scenario over the weak anthropic principle. The main advantage is that it leads to definite, falsifiable predictions. This is a general lesson. The Darwinian explanation is more vulnerable to falsification than its alternatives, because it produces a population that is highly improbable and full of unlikely structures and correlations. This is because there are as many "creators" as there are species, or universes—each is both created by its ancestors and creates its progeny. This structure is absent in a weak anthropic scenario in which there is a single creator. The logic is the same whether that single creator is an eternal primordial vacuum state or God. Theories with a single creator are untestable, because they lack all the detailed improbable structures and correlations that a system ordered by reproduction has.

To close, I would like to return to my opening remarks about the nature of law. The belief that beyond the temporary and contingent pageant of our world there is a transcendent reality outside of time has dominated Western thought since Plato. A Darwinian-style explanation, when applied at the level of the laws themselves, signals a decisive break with this ancient expectation and hope. For in a Darwinian universe, the laws may be no more than a description of a temporary equilibrium that is a step in an unstoppable evolutionary sequence. The Darwinian multiverse is falsifiable because it has lots of structure. But all that structure is contingent and time-bound. The population of universes is something that lives inside of time, and it is continually changing as new universes are created.

It is obvious, but must be said, that the Darwinian picture of cosmology I've presented here is only possible in a theory where the geometry of space is relational and dynamical. The processes of reproduction of universes literally grows space, as a system of relations continued in time beyond a bounce.

Einstein taught us that the geometry of space is not given by an eternal law; it is instead contingent and dynamical. Moreover, the geometry of space is relational: There is no meaning in Einstein's universe to "where" something "is," except in relation to something else. The same is true for time. Einstein's universe is an ever-evolving network of relations. Thus, Einstein and Darwin are part of the same great intellectual transformation. In the system devised by each, the old belief in a world where truth is defined only in relation to an eternal, Platonic absolute is replaced by a view of the universe as a relational, time-bound system in which all that exists must be created in processes of evolution.

I do not want to claim that all the implications of this shift are clear. There are deep and difficult questions to be resolved before we can comfortably see natural law as something that changes within time rather than something that exists eternally outside of time. For example, must there be a meta-law that guides the evolution of laws? For another, what implications does this have for the nature of time? Nevertheless, it appears more and more certain that this is where we are going. Cosmologists and theoretical physicists are talking less and less about what the law is and more and more about how some particular law, out of a vast list of possibilities, came to govern the part of the universe in which we live.

If Charles Sanders Peirce, more than 100 years ago, anticipated that Darwinian thought would be applicable at the level of the laws of nature, we can ask whether any contemporary philosopher has advice for how to carry on, once we absorb this insight. Here is a quote taken from a contemporary philosopher, Roberto Mangabeira Unger, who in 1987 anticipated the direction of contemporary cosmological theory.

> You can trace properties of the present universe back to properties it must have had at the beginning. But you cannot show that these are the only properties that the universe might have had. . . . Earlier or later universes might have had entirely different laws. . . . To state the laws of nature is not to describe or explain all possible histories of all possible universes. Only a relative distinction exists between law-like explanation and narration of a one-time historical sequence. . . .
>
> If you are asked what you mean by the necessity of the laws of nature (that is to say, by the necessity of the most necessary relations), you can legitimately respond only by laying out the substance of your cosmological and other scientific ideas. People who appeal to fixed conceptions of necessity, contingency and possibility are simply confused.

He could be talking about biology, but, as we have seen, he could equally well be talking about contemporary physics and cosmology. In fact, this quote is taken from a book called *Social Theory*. That the same kind of thinking is going on in all these realms shows us both the continuing influence of Darwin's thought and the possibility that we have yet to fully realize all of its implications.

LEE SMOLIN is a founding member and research physicist at the Perimeter Institute for Theoretical Physics in Waterloo, Ontario. A prominent contributor to the subject of quantum gravity, he is also the author of *The Life of the Cosmos* and *Three Roads to Quantum Gravity*.

STUART A. KAUFFMAN

Intelligent Design, Science Or Not?

I AIM TO PRESENT a rather nuanced view of intelligent design. I will argue that intelligent design is, at best, hardly science, and that insofar as it can be construed as science, the evidence that has accumulated runs strongly against it. Thus, it is obvious that intelligent design should in no manner be taught in schools as a creditable alternative to Darwinian evolution. This is, were it needed, made evident by the passionate support of intelligent design by the religious right, whose suspicions as to the identity of the unspoken designer are all too obvious. No other body of ill-posed science, with no supporting data, has been shoved and shoveled into the American school system. At base, the current pressure is scandalous.

Nevertheless, there is a core of "could be" science in intelligent design which must be acknowledged. Let us examine the case. Suppose some cosmic signal were picked up by SETI [the Search for Extra-Terrestrial Intelligence], and scientists, after examining the signal, announced that it was no mere cosmic chatter but in fact an organized signal. The question of the criteria that the scientists employed to affirm that the "signal" reflected an organized pattern emitted by some distant civilization would be a matter of serious and, I hasten to add, legitimate debate. That is to say, we can accept as a legitimate scientific question the issue of whether reception of word from ET really reflects ET's hand on the far-off switch, or merely the stochastic behavior of the interstellar medium at that observational wavelength.

In short, "Is it designed or not?" can be a legitimate and deep scientific question.

DARWINIAN PREADAPTATIONS AND INTELLIGENT DESIGN

Now let us turn to the biosphere. For virtually all biologists, the majestic unfolding of the biosphere over the past 3.8 billion years from the presumed origin of life on Earth is nothing short of stunning. Indeed, this propagating organization of process is as yet poorly articulated by contemporary biology and worthy of the deepest scientific interest.

Enter intelligent design and its moniker, "irreducible complexity." Here the keepsake example is the flagellar motor, a complex molecular assembly with an electric motor consisting of rotor, stator, and ancillary "machinery." ID argues, to state it briefly, that the motor is so complex, with so many parts each necessary for the adequate functioning of the whole motor, that no evolutionary procedure in assembling the parts could have attained the whole. Now, the standard informed evolutionary biologist, armed with Darwin, has a well-formulated response, given in terms of Darwinian preadaptations.

First, what is a Darwinian preadaptation? Consider the heart. Its function is to pump blood. But the heart makes heart sounds. These sounds are not the function of the heart. That is, the function of the heart is a *subset* of its causal consequences, and cannot be analyzed except in the context of the whole organism in its environment, life cycle, and selective regime. This fact has an important bearing, worth noting in passing, on the adequacy of reductionism, which, I argue, fails here. Let the reductionist deduce happily upward from a theory of everything to the heart. The physicist who does so can, in principle, state all the causal properties of the heart. However, he or she has no means of picking out the pumping of blood as the relevant feature of the heart for which the heart was selected, without conjuring organisms as causal agents in their own right and appealing to evolution and natural selection. The physicist must become a biologist.

So much for a thrust against reductionism. What is a Darwinian preadaptation? Darwin noted that a part of an organism might have causal consequences that were not of selective significance in the organism's current environment but in a different environment might come to be of selective significance. Thus, the swim bladder of an early fish evolved into the lung. The essential point here is that, with the Darwinian preadaptation, a new function has arisen in the biosphere. A further point of great interest is that it is not possible to *finitely prestate* all Darwinian preadaptations of, say, humans. This rests in part on the systematic vagueness of what constitutes an "environment." We cannot even get started on the enterprise of listing all possible environments. This in turn, means that the biosphere's evolution cannot be causally predicted. Much follows from this fact, which I have discussed in a recent book called *Investigations*. Among the implications is that the evolution of the biosphere cannot be given by predictions of changes in a prestated set of variables—for, via preadaptations, novel variables and functions (lungs and wings) persistently arise.

Second, preadaptations render probability statements based on the frequency interpretation of probabilities useless in the context of the evolution of the biosphere. Consider tossing a fair coin 10,000 times. It will come up heads about 5,000 times. But note that we can prestate what all the possible combinations of 10,000 heads and tails are—we can prestate the configuration space. This is just what we cannot do with respect to Darwinian preadaptations. We cannot prestate the configuration space of the causally relevant features of organisms. Thus we cannot use familiar frequency-based probabilities to describe the evolution of the biosphere. This bears on the argument of intelligent design, which is based on the familiar frequency interpretation of probabilities. Thus, to state that a given organ is so improbable that it requires design is just ill founded. The argument uses standard probability, which does not apply to the evolution of the biosphere.

Now back to the flagellar motor. The good Darwinian argues: "You see, the flagellar motor is the consequence of a long sequence of Darwinian adaptations. Initially, there was a pair of molecules that were selected to do *A*, but they were preadapted to do *B* in the presence of another molecule. In turn, *B* was preadapted to do C in the presence of another molecule. Ultimately a sequence of such Darwinian preadaptations led to the predecessor of the flagellar motor, which was preadapted to be such a motor if only a single further change was made. So"—smiles the biologist—"the flagellar motor was never selected *ab initio*—from nothing—to become a flagellar motor at all. Rather a sequence of sensible small steps, with persistent changes in functionality, led bit by bit to the flagellar motor. Thereafter, Darwinian normal adaptations refined the motor. So much for irreducible complexity!"

Is the standard Darwinian biologist right? Well, in some cases we have actually found intervening structures with functions that are not those of the final organ or structure. The aforementioned swim bladder of the early fish is an example. So in a reasonable number of cases, a reasonable number of intermediate steps with functions different from that of the final structure have in fact been found. This does not prove that we shall find all the intermediate structures in all cases, but such successes lend credence to the belief of virtually all biologists that such intermediate cases did exist and that for some, evidence may well be found.

Now back to the intelligent-design advocate. Can he or she make a testable prediction? Interestingly, the answer is Yes. The intelligent-design advocate must predict that in *no case* will such intermediate forms with diverse functionalities be found. After all, each such case of intermediate forms is evidence against irreducible complexity demanding a Designer. Taking ID as possibly science, then, since it can make testable predictions, what is its status? Well, we do have intermediate cases of sequences of Darwinian preadaptations leading to complex whole structures. These cases clearly count as disconfirming evidence to the universal claim of the ID scientist that no such cases will be found.

It follows from this that if we grant the intelligent-design advocates the maximum latitude to attain the status of a science, then intelligent design stands entirely without empirical support. I should add here that if the intelligent-design advocates refuse to take the prediction of the absence of signs of sequences of Darwinian preadaptations as one that is clearly implied by intelligent design, then they must give up the claim to be practicing science.

Of course, it does not follow that the Darwinians will always succeed in finding the intermediate preadaptation sequences. But science rests, as the philosopher W. V. Quine noted, on a rich web of interconnected laws, theories, and facts. In the absence of evidence for all cases of sequences of Darwinian preadaptations, the web of science as a whole overwhelmingly stands in favor of the Darwinian thesis. To hold the opposite, we would have to give up fundamental aspects of physics, chemistry, biology, geology, and so on. The web of scientific understanding would be torn asunder.

SELF-ORGANIZATION AND INTELLIGENT DESIGN

Beyond the issues raised in the foregoing, neither Darwinians nor proponents of intelligent design reckon with the growing evidence for profound self-organization that appears to play a fundamental role in evolution. Not all complexity requires astonishing selection. I discuss two cases.

The origin of life remains a mystery. There are several alternative approaches to the presumed origin of life on Earth, or on a nearby planet and thence transported to Earth by, say, ejecta. One approach is based on the hope of finding DNA able to replicate in the absence of a protein enzyme. This effort has failed for forty years but may yet succeed.

Second, there is the RNA world—the hypothesis that, at some early stage, life consisted of template-replicating RNA sequences, where RNA, acting as a ribozyme, was able to perform the function of a polymerase and copy RNA molecules, including itself. Some experimental evidence lies along the pathway to evolving such a polymerase. The hypothesis has merit.

Third, a number of workers have pointed to the ability of lipids to form closed bilipid vessicles called liposomes, much like the cell membrane in structure, and the ability of such liposomes to grow and divide. It may be that liposomes provide the most natural enclosure for complex molecular systems that carry out metabolism and synthesis of more complex molecules.

Fourth, I have investigated the hypothesis that complex chemical polymer systems, of a sufficient diversity, spontaneously self-organize into collectively autocatalytic polymer systems in which each polymer has its formation catalyzed by some polymer in the system. Here it is important to note that small collectively autocatalytic peptide systems devoid of DNA or RNA have been made experimentally. Thus, collectively autocatalytic sets are established. It remains to show that they can easily form in complex chemical systems.

Until a decade ago, I thought that the onset of self-reproduction was sufficient for the origin of life. Since then, I have been persuaded by my friends that confinement in something like a liposome is necessary as well. Indeed, one would like to show the self-organization of self-reproducing polymer systems that grafted into themselves a metabolism linking exergonic and endergonic (that is, spontaneous and nonspontaneous) chemical processes, and (as argued in *Investigations*) performing work cycles.

In summary, experimental approaches to the origin of life itself are well under way. We can hope in several decades to create life *ab initio*. Such a triumph would not, of course, establish the historical route to life's origin, but it would place the origin of life fully in the scientific rather than intelligent-design arena.

The second case of self-organization I will briefly describe concerns the control of cell differentiation from the fertilized egg to the adult. Human life starts as a zygote. This cell contains about 30,000 genes. The cell divides about fifty times, and changes, or differentiates, into about 265 cell types by histological critieria. Cell types differ because different subsets of genes are active in different cells, making different subsets of RNA and proteins. It has been known for sixty years that the protein made by one gene can bind next to another gene and turn it on or off with respect to making its RNA and protein. Thus, some complex genetic regulatory network orchestrates the differentiation that occurs.

It has also long been known that complex networks of model genes can spontaneously exhibit enormous order—order that appears sufficient to account for the order of cell differentiation and ontogeny. In a rough analogy, consider a mountainous region with many lakes, each draining streams from its own drainage basin. Nonlinear model genetic networks have the property that, if released from an initial state, or combination of gene activities, they will follow a trajectory through the set of states, and flow (like a stream) to a recurrent pattern of gene activity (like the lake). The recurrent pattern of activity is called an "attractor." A central hypothesis is that cell types correspond to such attractors (or lakes), while the various attractors correspond to the various cell types in the organism.

Stunningly, even randomly constructed model networks, subject to very simple constraints on the distribution of the number of regulatory inputs per gene and on the control rules governing the activities of genes as a function of the levels of their inputs, suffice to yield astonishing order. This is the self-organization of which I spoke. The order seen in the differentiation of cell types is not highly improbable but self-organized: "Order for free." As such, it cuts powerfully against the intelligent-design case.

Evolution needs to be recast as a marriage of self-organization and selection. Importantly, model networks behave in two broad regimes, ordered and chaotic, with a critical phase transition between them. Good theoretical work shows that the most complex yet controllable behavior occurs in critical networks. Thus it is deeply interesting that experimental evidence is just beginning to accumulate showing that prokaryotic and eukaryotic cells from distinct phyla all seem to be critical. Now, self-organization alone would not yield critical networks; rather, the marriage of self-organization (giving ordered networks "for free") and selection for the ability to coordinate complex tasks would plausibly yield critical cellular networks.

Based on self-organization as a new source of order in biology, and Darwinian preadaptations that render standard probability calculations meaningless and supply stepping-stones to complex novel functions by a sequence of simpler but different functions—plus the fact that such sequences have repeatedly been found and must be denied by the advocates of intelligent design—intelligent design, on the most liberal interpretation, fails to pass muster as creditable science.

Thus, in summary, intelligent design could be science: Were that true, it would be the most astonishing scientific conclusion of the past 350 years of modern science. To achieve stature and displace the contemporary coherent web of science would require overwhelming evidence. Where evidence of Darwinian preadaptations are known, they stand against intelligent design. Thus, were we to conjure with intelligent design as possible science, it stands without support.

And therefore, intelligent design should surely not be taught as "another theory" worthy of consideration. It is not logically impossible that intelligent design is correct, and America should be loath to give it credence.

STUART A. KAUFFMAN, the director of the Institute for Biocomplexity and Informatics, University of Calgary, is a theoretical biologist who studies the origins of life and of molecular organization. He is the author of *At Home in the Universe: The Search for the Laws of Self-Organization.*

SETH LLOYD

How Smart Is the Universe?

THE POPE RECENTLY called the universe an "intelligent project." Benedict's benediction of the universe's intelligence begs the question: Just how smart is the universe? Smart enough to evolve stars, planets, and life without any external coaching? Clearly, the universe is gifted, but what is its IQ?

Remarkably, the raw intelligence of the universe can be measured. We can't ask the universe to take the Stanford-Binet test or the SATs (unless it decides to apply to Harvard), but we can apply to the universe the same measures we apply to the raw information-processing power of computers. When buying a computer, we look at benchmarks like clock speed and memory space. Those same benchmarks can be applied to the computational capacity of the universe as a whole.[1] The universe can perform approximately 10^{105} elementary operations per second on about 10^{90} bits. That's a million billion billion billion billion billion billion billion billion billion billion billion operations

[1] For an overview of this application, see my article in *Physical Review Letters* 88 (2002), pp. 237–901.

per second on a billion billion billion billion billion billion billion billion billion billion bits. Over the 13.8 billion years since the Big Bang, the universe has performed around 10^{122} operations. The universe is pretty smart. Whether it is smart enough to evolve life without outside help will be discussed below.

BITS AND ATOMS

Let's look more closely at how the universe processes information. The number of operations performed by the universe and the number of bits available are big numbers (though not infinite). What do they mean? What does it mean for the universe to perform an elementary operation on a bit?

A bit, or "binary digit," is the smallest possible chunk of information. It represents the distinction between two possible alternatives, such as Yes or No, True or False, On or Off, 0 or 1. In a conventional computer, a bit is an electronic device, such as a transistorized switch, that can exist in one of two states: switch open ("0") or switch closed ("1"). The use of switches to register bits is a technological convenience; any physical system that can exist in two states registers a bit. For example, on a computer's hard drive, a bit is registered by a tiny volume (or "domain") of magnetic particles: When the north pole of this domain points up, that registers a 0; when it points down, that registers a 1. An elementary logical operation in a computer is just a bit flip. When you close the switch (or reverse the orientation of the north pole of the magnetic domain), the bit flips. In the past fifty years, the size of the devices that register bits in computers has gone down by a factor of a million: The vacuum tubes that made up the switches in electronic computers of the 1950s were a few inches in length; the transistors that make up the switches in computers of 2006 are a few thousand atoms across. Nor is there any reason to stop there. Over the past ten years, scientists and engineers have constructed quantum computers that store bits on individual atoms—one atom, one bit.

Atoms register bits the same way the magnetic bits in a computer's hard drive do. Each proton in an atom's nucleus is a tiny magnet; its north pole can point either up (0) or down (1). When a proton's magnet flips direction, its bit flips. Quantum computers operate by using atoms to register bits and by flipping those bits in order to perform elementary logic operations. But the atoms in quantum computers aren't the only atoms registering and flipping bits. *Every atom and elementary particle in the universe registers and processes information.*

The assertion of the intrinsic information-processing ability of atoms might at first sound like something out of *The Matrix*, or possibly only the twenty-second century. In fact, the information-processing capacity of atoms was well established by the end of the nineteenth. Indeed, the very formulas that mathematicians and computer scientists use to measure information today were developed starting in the 1860s by physicists who were trying to describe the motion of atoms. Scientists such as James Clerk Maxwell in Edinburgh and Cambridge, Ludwig Boltzmann in Vienna, and Josiah Willard Gibbs at Yale derived the mathematical expressions for the number of bits registered by the microscopic jiggling of atoms. Every atom registers a definite number of bits of information, they discovered, and whenever two atoms collide, those bits are transformed and processed. Long before the invention of digital computers, scientists had discovered that the universe computes. The numbers given here for the computational capacity of the universe were derived using simple, well-known techniques of physics to count the number of bits and bit flips in the universe as a whole since its beginning.

WHAT IS THE UNIVERSE COMPUTING?

We know how many flips the universe has so far performed on how many bits. So what? What is the universe computing? Maxwell, Boltzmann, and Gibbs would tell us that most of those bits reside in the random jigglings of atoms and most of those bit flips take place when atoms collide. Most bits are humble. But some bits lead more interesting lives. Every time a neuron fires in your brain, for example, it lets loose a torrent of bits. The cascade of bits in neural signals is the information processing that underlies your thoughts. The billions of bits of genetic information locked in the nuclei of your cells govern the minute chemical processes that constitute life, and (at least from the perspective of bits) sex is a glorious burst of information processing designed to pass on and transform that genetic information.

Words, both spoken and written, are a particularly powerful form of information, because they arise as part of a language. When you talk to me, your vocal cords vibrate and set up waves that travel through the air. The waves themselves consist of tiny displacements of atoms, leading to peaks of high pressure and troughs of low pressure. When they hit my ear, those waves set up vibrations in my eardrum and the small bones of the ear. Processed and transformed into electric signals, your words enter my brain and excite thought, and a response, in the form of words that I speak back to you. Although at every point along their path your words are registered by the motion of atoms and electrons, the significance of the bits they carry is far greater than the random motions of molecules in the air generating that slight hiss we hear in moments of otherwise perfect silence. The significance of words comes from their ability to refer to things—snow, air, the river—and from the ability of language to arrange words into arbitrarily complex forms of reference.

Human language is a recent development in the history of life on Earth. Life originated almost 4 billion years ago, whereas the fossil record and genetic evidence suggest that language originated between 50,000 and 100,000 years ago. The ability to speak and reason allowed humans to manage their environment, to create complex social structures, to become, in short, *Homo sapiens*.

Noam Chomsky argues that human language comes out of a basic conceptual development called recursion. Recursion is a process in which two things are put together to produce a third. For example, taking "I" together with "see" yields "I see." "I" is a concept ("myself") and "see" is an action; putting them together yields a sentence that combines aspects of "I" and "see." There are two important features of recursion: First, the two original things that are put together remain intact; second, the things produced by recursion can themselves be put together to produce new things. "You see" and "I see" can be put together to yield "You see I see." (Followed by "I see you see I see," and so on.) Recursion is the basis not only of human language but also of computer languages. Computer languages instruct computers to perform tasks (word processing, number crunching, game playing) that are built up recursively, out of simple underlying logic operations (AND, OR, NOT, COPY, and so on). Recursion is a simple process that is potentially very powerful. Prior to the development of recursion, human beings were just another group of primates. Now, as a result of the successes stemming from language, we are the first species in a billion years to produce global climate change.

While human language is a type of information processing, of which humans are justifiably proud, it is by no means the most spectacular information processing in Earth's history. Sexual reproduction also contains recursive elements: Two creatures combine their genes to produce a third. Life itself can be thought of as recursion: A living system interacts with its environment to produce another living system. Like language, life and sex are open-ended in the variety of combinations they can produce.

Even at its most microscopic levels, the universe processes information in nontrivial ways. Two species of chemicals will catalyze the production of a third without the original species themselves being consumed. At the level of elementary particles, a photon and an electron can interact to produce a second photon. (This is how a laser works.) At bottom, the language that governs how information is processed consists of the laws of physics themselves.

INFORMATION PROCESSING AND NATURAL SELECTION

Now that we know how smart the universe is, and what it is computing, let's look at the implications of the computational capacity of the universe for the origins of life and for natural selection. At every level, the universe is processing information. The form that information processing takes at larger scales is determined by the computation at smaller scales. The way in which elementary particles register and flip bits determines how atoms process information. The way in which atoms process information determines how molecules behave. The way in which molecules behave determines the structure of life. And so on, up to dogs, humans, politics, and society.

The more microscopic the form of information processing, the longer it has been going on. Information processing at the scale of elementary particles began with the Big Bang itself. Information processing at the level of atoms began some hundreds of thousands of years later, when the universe had cooled sufficiently to allow atoms to form, and molecular information processing began once matter had cooled and become sufficiently dense for significant amounts of molecules to form and interact chemically.

For life to begin, some chemical had to be able both to catalyze its own production and exhibit chemical variation at the same time. Once self-reproduction and variation were in place, natural selection could begin. While we don't know in detail how life began, the ability to catalyze one's own production is common in chemistry, even for quite simple molecules. The origins of life did not necessarily require complex chemistry of the sort exhibited by life today. So the universe had to be smart to produce life—that is, it had to be able to process information in the first place—but it did not require extra coaching. Only the simplest kind of chemistry was required for the original self-reproducing molecules to arise.

If those proto-life-forms were simple, why haven't biologists conclusively identified them yet? The reason is straightforward. Even if the origins of life involved simple chemistry, there are many, many possible simple chemical combinations from which life could have arisen. In particular, there are many more possible candidates for proto-life than there are graduate students to try them out. In contrast to the biological research community, however, the universe as a whole possesses a huge diversity of laboratories, in the form of planets, and planets possess a huge diversity of chemical conditions and environments. In addition, the universe has had the leisure of billions of years to try out different chemical combinations until one capable of self-reproduction with variation arose. Not only is the universe smart, it is plenty smart enough to discover life on its own without supernatural help. If human biologists keep trying, maybe they too will hit upon the same fruitful chemical combination that the universe found billions of years ago.

NO FREE LUNCH?

Even if the universe is capable of creating proto–living beings that reproduce with variation, it may seem at first implausible that this ability alone is enough to set the process of life and natural selection rolling. However, we have strong computational evidence that it *is* enough. Computer scientists and biologists can program artificial life-forms into computers to see how they evolve. These investigations have demonstrated that as soon as artificial organisms are able to reproduce with variation, they construct a complex artificial ecosystem that includes viruses, the exchange of genetic material, and artificial biological "arms races," just as real ecosystems do.[2] The preliminary successes of such artificial-life investigations do not mean we have solved the problem of how life evolved on Earth, but it does supply strong evidence that simple reproduction together with variation is enough for it to have

[2] See for example the "Tierra" system developed by University of Oklahoma biologist Tom Ray, in which self-replicating machine code programs evolve by natural selection.

begun. Exactly how life came into existence is still a mystery; that it should have come into existence at all is not. Because of the universe's information-processing power and diversity, it was virtually certain to hit upon life sooner or later.

Intelligent-design proponent William Dembski argues, however, that the problem of finding the proper combination of ingredients for proto-life is "too hard" for the universe. In support of his position, Dembski invokes the no-free-lunch theorem developed by the computer scientists David Wolpert and William Macready, which states that an algorithm good at finding patterns in one kind of environment is necessarily bad at finding patterns in another kind of environment. Consequently, no chemical computation on its own, Dembski claims, has the computational ability to discover a form of proto-life that can adapt to varying environments.

Dembski's argument involves a misunderstanding of the situations in which the no-free-lunch theorem can be applied. First, proto-life on Earth need adapt only to terrestrial environments, which, while diverse, are much less so than those required to apply the no-free-lunch theorem. Rather, that theorem explains why life on Earth might be poorly adapted to living on Venus; an organism such as a fern that has evolved the ability to scrounge a free lunch on Earth out of air alone would rapidly shrivel and die in the burning hot Venusian atmosphere. Second, the environment that proto-life faces is not random (as required by the Wolpert-Macready theorem) but is instead highly ordered, having coevolved with that proto-life itself. Much more relevant for the origins of life than the no-free-lunch theorem is the following sequence of theorems, which I dub the free-lunch theorems.

Free-lunch theorem 1: Any system that is computationally universal has a nonzero probability of producing as a subsystem any computable structure, including the laws of physics, all computable mathematical structures, and so forth.

Free-lunch theorem 2: Any system that is computationally universal has a nonzero probability of producing subsystems that reproduce themselves with variation.

Free-lunch theorem 3: The reproducing subsystems of theorem 2 have a nonzero chance of eventually producing subsystems capable of recursion.

Free-lunch theorem 4: If the original system of theorems (1–3) possesses an intrinsic source of variation, then any given computable structure will arise somewhere in the original system with probability equal to one—that is, for certain—and so will reproducing subsystems and subsystems capable of recursion.[3]

The application of the free-lunch theorems to our universe is clear. We have incontrovertible empirical evidence that our universe is capable of universal computation: I am writing this on a computer. In addition, our universe does have an intrinsic source of variation: The laws of quantum mechanics, which, together with classical chaos, continually inject new random bits of variation into our world. At bottom, these random bits from the quantum realm are the source of all the variation we see around us, from the positions of stars in the sky to the mutations in our genes. Consequently, the free-lunch theorems confirm the ability of the universe to produce life and the complex world we see

[3] The proof of these theorems is a straightforward consequence of Kolmogorov's theory of probability based on algorithmic information.

around us. As my MIT colleague the cosmologist Alan Guth has pointed out, the universe is the ultimate free lunch.

THE NATURAL SELECTION OF SCIENTIFIC THEORIES

So far, we have been talking about natural selection in the past. Let's finish by talking about natural selection in the future. In particular, let's use the ideas developed here to investigate the natural selection of scientific theories.

The computational picture of the universe consists of a sequence of information-processing revolutions, each of which arises from the previous one. Elementary particles explore all ways of colliding with one another until eventually, in the right circumstances, they form atoms, a form of matter selected for because of its stability. Atoms collide in every possible way until they form a wide variety of molecules, each selected for by the local concentration of atoms together with the laws of chemistry. Molecules, in turn, explore ever more complicated chemical reactions until they form a molecule capable of catalyzing its own production together with variation in its form: Such a form of proto-life is selected for merely by its ability to reproduce and adapt to different environmental conditions. Because of its ability to adapt to new surroundings, life explores a vast space of possible beings, until it arrives first at sexual reproduction and then at language.

Once language arises—at least, human language—we can know much more about what kinds of behaviors are selected for, simply because oral and written history tell us what has succeeded and what has not. Social organization, if well done, endows the organizers with a greater chance of reproducing and passing on not just their genes but also verbal information about how to perpetuate that social organization. Such social innovations have historically led to great benefits for the people propagating them and sometimes to misery and even death for those opposing them.

Science is a particularly powerful form of information processing based on language. Scientific knowledge consists of that information whose accuracy has been confirmed repeatedly by observation and experiment. Science progresses in exactly the same way as all previous progress in information processing: A great variation of ideas, based on combinations of previous ideas, is put forward and tested against the actual world. Simply because an idea has been proposed by a scientist does not make it scientific knowledge; the vast majority of ideas proposed by scientists turn out to be wrong. That is as it should be. We cannot expect all our ideas to be right, no matter how attached to them we are for personal or ideological reasons. To construct a scientific idea that is even potentially correct, one must ensure that observation or further calculation could prove it incorrect. Even then, decades or even centuries may pass before even the best-confirmed scientific ideas become firmly established. For example, witness the current attacks on Darwin's theory of evolution by natural selection despite the fact that this is one of the most thoroughly confirmed scientific theories ever, with hundreds of billions of pieces of confirming experimental evidence. (Ironically, witness the current lack of attacks on string theory, a theory of fundamental processes that boasts hardly any confirming observational evidence.) Most of the ideas proposed by scientists are wrong, but a few ideas are right enough to survive the harsh environment of experiment and scientific criticism. Those ideas that survive are what we call "science."

Scientific knowledge is by definition resilient. In societies where government or religion has tried to replace it with ideologically inspired fictions, scientists and nonscientists alike have resisted. Scientific lies can fool some of the people some of the time (even to the extent of being published in reputable journals), but exactly because scientific ideas are designed to be tested, in the end scientific lies fool no one. The universe is scientific. The universe has successfully designed all sorts of beautiful structures by trial and error, selected by natural laws, since the beginning of time. Its use of natural selection as the primary form of design is the primary confirmation of how smart the universe is. Attempts to circumvent the process of natural selection not only of life but of ideas, are unlikely to succeed. The mills of science grind slowly, but they grind exceedingly fine. "Intelligent design" is not grist for that mill: It is chaff, and it will blow away. Let us hope that it blows away before it has too severely damaged the minds of the upcoming generation.

SETH LLOYD, a physicist and professor of quantum engineering at MIT, is a leading researcher in quantum computation and quantum communication. He is the author of *Programming the Universe: A Quantum Computer Scientist Takes on the Cosmos.*

LISA RANDALL

Designing Words

SCIENCE INVOLVES A SET of falsifiable assumptions and a framework for predicting and understanding observable phenomena. It arises from an amalgam of observation, mathematics, intuition, and experiment. Hypotheses that contradict facts or aren't testable are not science. However, we don't always immediately know which ideas can be tested or which hypotheses will agree with experimental results. Advances often happen in fits and starts, so that the progression of science becomes clear only after evidence and theory have settled down. But once new phenomena are understood, different areas of science can converge in surprising ways, since, after all, the various areas of science apply to the same world.

In the original (1859) version of Darwin's *On the Origin of Species by Means of Natural Selection*, he discussed his evidence that the earth was older than 300 million years. He had looked at the erosion of the Weald, a lowland in the south of England, and deduced its minimum age. However, in 1862, William Thomson (later Lord Kelvin) had other ideas and used physical principles to deduce the sun's age. He asked how long the sun could have sustained itself via combustion, then the only known way for the sun to burn and emit heat and light—and deduced that it was a mere 30 million years old.

William Thomson was the leading physicist of his time. Darwin didn't understand why the results based on Thomson's physics calculations didn't agree with his own conclusions but, recognizing that there was a problem, he removed the discussion of the Weald from later editions of his book.

But it turned out that Thomson was wrong. The sun has enough energy to burn for much longer than he had anticipated—billions of years, in fact. At the time, no one knew about the weak and strong nuclear forces, which provide the primary mechanism for fueling the sun. Once they were discovered, the discrepancy between Thomson's and Darwin's conclusions was resolved. Darwin's observations had not deceived him. The theory on which Thomson had based his 1862 result was correct, but it wasn't the whole story.

As a theory progresses, it is uncertain in the beginning and is later resolved or absorbed into a more comprehensive theory. Scientific quandaries or contradictions are frequently a sign that a theory is incomplete, not necessarily wrong. Gaps in our understanding serve as incentives for questions and further scientific advances, not for abandoning the scientific enterprise. The biggest difference between science and a belief system is that if we find contradictions in our theories, or if more precise measurements contradict them, we will modify or abandon them and move on, while holding on to those aspects that have proved successful.

Unfortunately for the communication of scientific ideas to the public, neither the difference between "science" and a "belief system" nor the idea of "effective theories" that work up to a point and then are absorbed are always properly understood. But those are only a couple of examples of the many misunderstandings about science. Ambiguous scientific terminology is the source of some other misconceptions. Scientists often use colloquial terminology to which they assign a specific meaning that can't be fathomed without proper training. The physics term "uncertainty principle," for instance, is frequently interpreted as a limitation on observers and their ability to make measurements. But it is not about intrinsic limitations on any one particular measurement; it is about the inability to precisely measure particular pairs of quantities simultaneously.

Such ambiguities can easily distort the implications of a theory or result. These confusions are unfortunate but are generally harmless. However, ambiguous word choices become especially precarious in the context of the current debate over evolutionary theory. As a physicist, I'm less affected by such discussions than my colleagues in biological fields. But an examination of the words used in discussions of evolution and some aspects of the history of the science elucidates some interesting aspects of the nature of science and how it is understood.

Even the word "evolution" itself is sometimes ambiguous—ironically, in part because of the evolution of that word over time. Perhaps one of the most interesting criticisms of evolution by natural selection in Darwin's era came from the Swiss-American geologist, glaciologist, and zoologist Louis Agassiz: Agassiz didn't deny that evolution occurs in nature, but his idea of evolution was that it entailed the preordained unfolding of a plan. Today the word "evolution" is sometimes incorrectly or too narrowly applied, lending false credence to the critics of the theory.

Probably the greatest confusion in the use of the term today arises from the failure to distinguish between the fact of evolution—that is, that heritable changes have occurred in a population over generations—and the mechanism by which evolution occurs. The evolutionary mechanism as understood by Darwin was natural selection—the differential survivals of organisms depending on the degree of their fitness to their environment. But there are alternative hypotheses at work now—genetic "drift" and symbiogenetic modification, among others. As cellular and molecular biology develop, we are learning more about the processes by which evolution occurs. Thus the *mechanism* of evolution is a subject of ongoing research, but the *fact* of evolution has been pretty definitively established.

We can readily find evidence that demonstrates that modern species are evolving and that closely related species share a common ancestor. We now can say with confidence that Earth is more than 3.6 billion years old and that cellular life has existed for more than half that time, while organized multicellular life has existed for at least 800 million years. There were no birds before 250 million years ago, and major life-forms from that time no longer exist. Yet we can identify enough common elements to determine that living creatures today evolved from living creatures of the past.

The evidence for the fact of evolution includes the fossil record, which reveals gross characteristics that overlap but are not identical and genetic frequency counts that are similar but not the same. There is concrete evidence for speciation—as evinced by Darwin's finches on the Galápagos Islands, for example, and also in the fossil record, where we can see systematic changes with a definite time progression in almost all cases. There is evidence for life-forms intermediate between fish, amphibians, reptiles, and mammals. Homologies—common features in the anatomy of different creatures—also support descent from common ancestry, and these conclusions are consistent with the fossil record. Other evidence for the fact of evolution comes from studying the distribution of species, which develop independently when, for geological or other reasons, species become isolated. And more species develop whenever the environment is more isolated and offers less competition from existing life.

But evidence today is not restricted to the fossil record or to biogeography. It also comes from molecular biology, genetics, cell biology, and embryology. Embryology gives evidence of relatedness of very different-seeming animals, both in form and in common genes. Molecular biology provides powerful evidence both in DNA and in proteins, which are similar but not exactly the same in many different organisms. Such results can be cross-checked with the other evidence for evolution, and the progressions that are deduced coincide. And we should not ignore the evolution we observe today, in the proliferation of antibiotic-resistant bacteria and insecticide-resistant bugs, for example.

Is this evidence that evolution is a fact? No scientific truth can be established with 100 percent certainty. But a fact for scientists is something so likely that it is established beyond reasonable doubt—and evolution fits this criterion. Imagine that someone who is hell-bent on the theory of intelligent design as an explanation for Earth's flora and fauna faces up to this evidence. What sort of designer would he or she envision? The only sort of designer who could have created all the useless appendages, the many extinct species, and the various intermediate states with their peculiar features would be one who tries out different possibilities and then keeps the ones that work. Perhaps the designer is imaginative, or even playful—but obviously not omniscient. Moreover, the designer would have to be mindful of the environment of the species that it creates. Not only do species adapt appropriately—particularly when a species gets transported to an isolated environment—but speciation occurs with much more abandon when fewer competing species are there to prevent this from happening. Finally, the designer has to be an evolving designer. There is a definite vector of time in the fossil record. It doesn't go backward. Once change happens, there is no turning back.

So what are we left with if we want to anthropomorphize the process by which evolution has occurred? We are left with an evolver, not a designer—one who gets bored or dissatisfied with changes and moves on. Whoever is responsible is just trying out the various possibilities. We don't have an intelligent designer (ID), we have a bungling consistent evolver (BCE). Or maybe an adaptive changer (AC). In fact, what we have in the most economical interpretation is, of course, evolution.

But what about the phrase "evolutionary theory?" Even a word as basic to science as "theory" causes confusion, providing a field day for advocates of intelligent design. By conflating a scientific theory with the colloquial use of the word, creationists instantly diminish the significance of science in general, and evolution's supporting scientific evidence in particular. Unlike most people, who use the word "theory" to describe a passing conjecture that they may well regard as suspect, scientists have very specific ideas in mind when they talk about theories. For scientists, theories entail a definite physical framework embodied in a set of fundamental assumptions about the world—and an economical framework that encompasses a wide variety of phenomena. A theory yields a specific set of equations and predictions—ones that are borne out by successful agreement with experimental results. We use the word, for example, when we talk about "the theory of general relativity" or "quantum field theory." Science isn't just about observations. If one refuses to attribute any causal relationship to connected phenomena, or to see connections that follow from some basic underlying assumptions about how things work, it's not science.

That doesn't mean that all theories are shown to be correct immediately. Even Einstein took the better part of a decade to develop the correct version of his theory of general relativity. But eventually theories are either proved correct or abandoned. Or a previously successful theory might be absorbed into another, more encompassing theory when new regimes of distance or energy or more detailed investigations become accessible.

The current debate over evolutionary theory dates all the way back back to Darwin. Agassiz, one of Darwin's most notable critics of the time, was not an evolutionary theorist, but (like Darwin) he was a born naturalist. Both he and Darwin recorded the details and distinguishing characteristics of animals as seen in their environment and in the fossil record. But Agassiz's approach to the Earth's flora and fauna was strictly empirical. He carefully observed nature and possessed an extensive knowledge of natural history but avoided the theoretical relations among what he found. (When a statue of Agassiz on the Stanford campus was toppled by the 1906 earthquake, the president of the university, David Starr Jordan, who was his former student, remarked that Agassiz had always fared better in the concrete than in the abstract.)

Agassiz's strength was in description and classification rather than interpretation. His early work on fishes more than doubled the number of all described fossil vertebrate species. However, he viewed species as abstract categories rather than material entities. And he considered Darwin's ideas about speciation as "a scientific mistake, untrue in its facts, unscientific in its method, and mischievous in its tendency." Part of the reason for Agassiz's doubt was the incompleteness of the fossil record at the time. Even Agassiz conceded that he "would have been a great fellow for evolution were it not for the breaks in the paleontological record." That record today is still imperfect but far more complete than it was in the late 1800s and provides solid evidence for Darwin's theory. Interestingly, both Agassiz and Darwinians observed similarities between the stages of embryonic growth and the geological succession of species. Darwin considered this evidence in favor of evolution, whereas Agassiz saw this as evidence that species remain fixed.

Agassiz acknowledged that he and Darwin had access to the same data. But they came to very different conclusions. Agassiz preferred to believe that species were more or less fixed, and he wanted to find evidence for this in the fossil record. Before Agassiz accepted Darwin's idea of "transmutation"—that is, the formation of new species from older species—he wanted to see concrete evidence in the fossil record that one "variant" predated another. His ideas weren't necessarily religious, but he had a philosophical, almost Platonic, desire to see perfection in the world. Even with a complete record, he might still have resisted, if he chose to, by seeing the evidence as part of a preordained plan. Scientists through the ages have often resisted the idea of the universe being imperfect and changing. Einstein faced the same disturbing fact when he worked out the consequences of his theory and found that the universe was expanding. At that time, there was no experimental evidence for an expanding universe, so Einstein made what he later confessed to be his "biggest blunder" and introduced a cosmological term to adjust this theory to make a static universe. But it turned out that the universe does evolve. It's not static.

How could Agassiz and other scientists look at the same scientific evidence and draw such different conclusions? Agassiz's objections address both the nature of observation and the conduct of science. He could observe patterns but not causation. He refused to acknowledge the causal relationships intrinsic in evolutionary theory, even when he had identified many of the links that suggested it.

Most people think of "seeing" and "observing" directly with their senses, but for scientists the words "see" and "observe" refer to much more indirect measurements, involving a train of theoretical logic by which we can interpret what is "seen." I do theoretical research on string theory and particle physics and I try to focus on aspects of those theories that we might someday experimentally test. My most recent research is about extra dimensions of space. Remarkably, we can potentially "observe" evidence of extra dimensions. But we won't observe in ways you probably think of when you hear that word. We won't reach out and touch extra dimensions with our fingertips or see them with our eyes. The evidence will consist of heavy particles known as Kaluza-Klein modes that travel in extradimensional space. If our theories correctly describe the world, there will be a precise enough link between such particles and extra dimensions (a link that will be experimentally demonstrated) to establish the validity of the extra dimensions. The advanced mathematical nature of the theory that describes these particles and dimensions and the indirectness of experiments to detect them don't make these observations any less real. The way scientists see today is mostly indirect. We see evidence consistent with a particular interpretation of events. Observations and theory are intertwined, because any observation at the border of understanding will need to be interpreted. This was one of the key insights that Einstein gave to Heisenberg which led to the latter's formulation of the uncertainty principle.

Agassiz saw the power of a Creator in nature's flora and

fauna. But he believed in a series of creations, destroyed by past catastrophic events such as an ice age. While he believed that the word of God was revealed in scripture, his views were not strictly scriptural: He also believed that the word of God was revealed in the laws of nature. And Agassiz was willing to accept an Earth older than is deduced from the Bible. Agassiz saw the deity as "a busy, imaginative, enthusiastic, benevolent God with a love of form and order, a concern for detail, an appetite for variety, a passion for progess, an eternal readiness to start fresh." (Interestingly, Agassiz's God sounds a lot like Agassiz himself.)

Agassiz was perfectly willing to develop a coherent scientific theory when it came to the natural history of the earth. He helped formulate the theory of glaciation and observed how it accounted for various geological phenomena that had formerly been attributed to the Noachian Flood. Darwin praised him for his work on glaciers and Charles Lyell, the founder of modern geology, heartily approved as well. Lyell did for geology what Darwin did for evolution, developing it into a widely accepted science. It's fascinating to me that Agassiz was willing to accept the modern geological view of Earth's evolution while refuting the modern view of how life evolved, since both of those views contradict the notion of a static, perfectly created world.

Ironically, Agassiz was responsible for establishing Harvard's Museum of Comparative Zoology in 1859 (the same year that Darwin published the *Origin of Species*), an institution that ended up providing some of the soundest evidence in support of Darwinian evolution. Agassiz donated his own collections to the new museum and was able to convince others to give him the means to purchase other collections. These collections of fossil invertebrates remain central to current research. The MCZ has proved to be one of the most valuable resources in studying speciation through the ages and has provided fodder for the likes of Stephen Jay Gould and E. O. Wilson. As Alfred Russel Wallace noted, "It is surely an anomaly that the naturalist who was most opposed to the theory of evolution should be the first to arrange his museum in such a way as best to illustrate that theory. . . ."

None of this is to say that discussions of evolutionary theory are over. We have yet to learn how the initial conditions for evolution came about. Evolutionary theory is in far better shape than, say, string theory, from the perspective of how firmly it has been established. String theory, at this point, is entirely untested experimentally and some physicists fear it will never be tested. But string theory, even though it is not fully understood, provides a conceptual or mathematical framework from which we can deduce consequences; it is not simply a hypothesis. String theory is motivated by gaps in the theoretical framework that inspire further questioning. And string theorists are scientists who accept the well-established body of knowledge that tells us that general relativity and quantum mechanics are correct in their tested regimes.

Evolutionary theory is incomplete as well, though it has been far better tested. For example, we don't have an argument based on natural selection for why we have twenty-three pairs of chromosomes. And natural selection can operate on different levels—the group level, the individual level, and the genetic level—and it is not always clear how that level is to be determined. Our discussions are complicated not only by the less-than-accurate knowledge of the mechanisms responsible for evolution but also by our inability to perform experiments to test the evolution of all possible species. Sciences like geology and astronomy face the same sort of experimental problem. The necessity of relying on an incomplete historical record can impede progress, since the necessary evidence or pointers may just not be observable, having been destroyed or obscured millions of years ago. Nonetheless, these sciences progress by examining the existing evidence and extrapolating known physical processes back into the past and out into the universe.

Evolution, which involves living organisms, presents us with a further hurdle: We don't understand life itself, or many of the essential underlying processes that led to it. But of course that's what makes it such an exciting field. As we better understand microbiology, we can apply what we learn to understand processes and changes that occurred in the distant past.

Evolutionary biology is a complex science, and many questions will remain for a long time. But scientific criticisms of evolution don't deny the preponderance of evidence supporting natural selection or the fact that animals have evolved from their ancestors. While scientists have to acknowledge the limitations of their current theory, skeptics have to realize that the way to progress is by scientifically addressing the missing elements, not by misleading those who don't understand the science or by ignoring the evidence. The current "debate" over what to teach in science class is just embarrassing.

Perhaps the biggest word problem in this debate over evolution isn't a scientific one at all: It's the whole idea of "design." Design is creation with a purpose. But scientists don't ask questions about purpose; they ask questions about consequences. Newton's laws don't tell us why a ball *should* follow a certain trajectory; it just tells us that it does and how to calculate it. Consider this exchange from a children's science book from 1836, titled *First Lessons on*

Natural Philosophy, in which the distinction between scientific and religious questions is made plain:

> "*When you throw a ball up in the air, what becomes of it?*
> It moves in the direction in which I threw it.
> *What makes the ball move on?*
> The power of my arm.
> *Do you make this power?*
> I do not; God made it."

For today's natural philosophers, it is so obvious that science doesn't answer certain kinds of questions that we don't even bother to ask them. The answers may or may not have something to do with religion, but that's just not relevant to science. Religion, if it plays a role, takes off where science ends. Some people restrict their application of religion to questions of purpose that don't contradict or overlap with science. But historically religion has applied to the natural world. Religion then is a moving target: As we understand more about the natural world, its role recedes to questions beyond the ones that science successfully addresses. However, either way, religion and science are separate. Perhaps the real question we should be asking is why anyone would deny science the right to address the questions it is designed for.

LISA RANDALL is a professor of physics at Harvard University and a pioneer in the field of extra dimensions. She is the author of *Warped Passages: Unraveling the Mysteries of the Universe's Hidden Dimensions*.

MARC D. HAUSER

Parental Guidance Required

I ASSUME THAT every parent and educator will agree that teachers of science should passionately convey the importance of learning and the excitement that comes from mastering a discipline, including its theories and discoveries. But like sports and board games with rules of play, there are rules for science education—rules that enable our children to absorb its ideas and methods of inquiry and avoid being fooled by fakes. We want our children to see science as a way of knowing, a methodology that encourages their curiosity and empowers them to detect such hoaxes as crop circles, Piltdown man, and the "Stone Age" Tasaday of Mindanao. We want our children to relish the excitement of discovery, to ponder the great unknowns, to dream of contributing one day to our knowledge of the world around us. And even if they never make a scientific contribution of their own, we want them to be educated consumers who understand why the discovery of a new physical force, molecule, fossil, species, brain function, or sociological process is important. Without this appreciation, ignorance will settle in—and with today's rapid advances in science and technology, ignorance is a handicap.

To personalize a bit: I have school-age children, and the recent debates about whether or not to teach intelligent design in the nation's schools have me deeply worried. I am concerned not only about what American children will be learning about evolutionary biology but about the effect of those debates on education in general.

If you're not worried, you should be. Let me explain by citing a particular circumstance—the Kansas Board of Education's August 2005 decision to change the definition of "science" to include the supernatural as a relevant domain of evidence—to illustrate a general distinction between what we *can* teach and what we *should* teach. This essay is about the "coulds" and "shoulds" of education. I will argue that we *could* allow a wholesale blurring of disciplinary boundaries when we teach, including not just objectively derived facts but also intuition, speculation, opinion, and first-person or anecdotal experience in every discussion and debate. But this would be a profound mistake. It would leave students in a state of utter confusion, ill-prepared to understand scientific, technological, or medical developments and equally helpless in pursuing, say, a religious life. They wouldn't understand why personal experience or introspection is insufficient evidence for the safety of a particular drug—nor why thousands of test trials cannot "prove" that a belief in God provides meaning to one's life.

A more satisfactory curriculum would have teachers maintain disciplinary integrity, so that each discipline's goals and methods were clearly explicated, while generating new courses at the interfaces of disciplines. The new courses would celebrate the excitement that comes from cutting across disciplines and preserve the signature styles of each, including their distinctive questions, methods, assumptions, and confirming evidence.

Consider a hypothetical high school curriculum. The opening course in mathematics is geometry. Suppose the math teacher opens with a discussion of Plato, the inscription on his academy ("Let no one ignorant of geometry enter"), and Socrates' quizzing in the *Meno* of the slave boy to bring out his inherent geometrical knowledge. Introducing history into a geometry course is not only legitimate but useful; it provides students with a sense of the earliest stages of knowledge and how understanding has progressed up to the present. But suppose the teacher continued in this way for the entire term, talking only about the Greeks' discoveries, including Pythagoras's theorem about triangles. Students would leave the course without being able to find congruent polygons, calculate the areas and volumes of polygons and polyhedrons, or understand the use of algebra in analytical geometry. In a geometry course, teachers should teach the mechanics of geometry, throwing in history only when it enriches a particular section, perhaps by showcasing prior controversies. A mathematics teacher could certainly give a course on the history of geometry or the philosophy of mathematics, full of rich details about the characters involved, their personal lives, and their disagreements. But this would be a different sort of course, not one that could prepare students for the follow-up course in trigonometry. I assume that this much is uncontroversial.

The opening science course in our hypothetical high school curriculum is physics. Paralleling the introduction to geometry, the physics teacher begins with some early history, focusing on Aristotle, Copernicus, Galileo, and Newton, especially their theories about the heavens and the relationship between Earth and the other planets. As the instructor points out, the Roman Catholic Church saw the Copernican and Galilean thesis as dangerous to faith, ultimately sentencing Galileo to a life in exile, banished from science. The instructor then moves to more modern wars between physics and religion, as well as the legendary comments by distinguished physicists about the role of a divine power in creating the universe. The instructor then ends the course with a discussion of Einstein's conviction that "everyone who is seriously involved in the pursuit of science becomes convinced that a spirit is manifest in the laws of the universe—a spirit vastly superior to that of man. . . . In this way the pursuit of science leads to a religious feeling of a special sort, which is indeed quite different from the religiosity of someone more naive." As the instructor notes, even Einstein was humbled by the immensity of the problems he confronted in explaining the universe, leaving a divine entity to explain the existing mysteries.

If my children were in this class, I would be thoroughly annoyed. And even if they were not my children, I would be annoyed by the approach and the disservice to general education. Although I would applaud the interdisciplinary flavor of the material as well as the instructor's use of history to inform our current understanding of physics, I would argue that such students were ill prepared for college physics, or even the follow-up high school courses in chemistry and biology. Without a basic understanding of motion, energy, force, magnetism, electric fields, waves, current, resistance, solids, and fluids, along with the basic theories and methods behind quantum, atomic, and nuclear physics, our children would fail not only to move on in science but to appreciate the general laws of the physical world.

As with the complementary mathematics course, I would be delighted to see a course on how physicists throughout history have brushed up against religion, both in their challenges to religious faith and their reliance on divine beings to account for the mysterious. This is a course that *can* be taught, but I would not want the school, or the state, to legislate that it *should* be taught. We should not offer this kind of course as an alternative to the material necessary for understanding the theory, methods, and discoveries of physicists. If it is an introductory course in physics, students must understand the relationship between theory and evidence, the assumptions underlying particular interpretations of evidence, and (to keep things alive) the nature of current controversies. But what counts as a controversy must be delineated with care, as we want students to distinguish between scientific challenges and sociopolitical ones. Experiments, observations, and theories that challenge the validity of a purported position in physics are all fair play. Religious beliefs, politically motivated ideologies, and incredulity are not, because they are operating on a different level—one of emotion and intuition. Needless to say, controversies within physics and the other sciences are fueled by emotion and often guided by intuition, but these are not the relevant means of adjudicating among conflicting positions.

The opening literature course in our high school *manqué* is "Introduction to the Bible." The instructor begins with a discussion of Charles Darwin, noting that he was a lifelong member of St. Mary's Church in Downe, England; served on their parish council; provided financial support for the parish's educational programs; and was a close friend of the vicar, the Reverend John Brodie Innes. The instructor then turns to a detailed analysis of Darwin's ideas on evolution—in particular, his theory of natural selection and the data he amassed over several decades in support of its primary tenets. From Darwin, the instructor moves to more recent discoveries in evolutionary biology, including the theoretical insights of William Hamilton, George Williams, Robert Trivers, and John Maynard Smith, as well as the methodological developments in molecular biology, animal behavior, neurobiology, and paleontology. These discussions are intertwined with an analysis of the evidence for our evolutionary past and the notion of human nature. The course ends with a brief comparison of biological origins in Genesis and Darwin's account in *On the Origin of Species*.

As a parent, I would, once again, be annoyed. Though many consider Darwin's *Origin of Species* an important piece of literature and certainly worthy of comparison with the Bible, the centerpiece for the course should presumably be a textual analysis of the latter. Whereas Darwin's writings may be relevant to a study of other accounts of our origins, the skills required for literary scholarship are entirely different from those involved in scientific scholarship. When a text like the Bible is subjected to literary criticism—as opposed to its more traditional role as a source of religious inspiration and guidance—the methods are different from those employed by a scientist scrutinizing the merits of a theory or piece of experimental evidence. I want students to learn the art of literary criticism and to distinguish it from other kinds of critical thinking. I want students to distinguish between the fictional artistry of Dan Brown's *Da Vinci Code* and scholarly treatments of the same history, such as Elaine Pagels' *Gnostic Gospels*. And I want our educational systems to distinguish between what high school teachers *can* teach and what they *should* teach in order to prepare our children to be intelligent grazers, digesting information gathered by scholars in the arts, humanities, and sciences.

Anyone who shares my diagnosis of this curriculum should be appalled by the recent policies inspired by supporters of intelligent design and accepted by the Kansas State board. Supporters of this movement are functionally harming their own children as well as the children of dissenting parents. The scientists who support evolution (by far the majority) are portrayed as fundamentalists, myopically focused on a singular thesis, immune to conflict, and unwilling to entertain supernatural "evidence" as evidence—in short, approaching the old Hollywood caricature of the evil scientist. Parents need to be equipped with information to fight back as intelligent consumers. Here's some equipment.

Let's return to our hypothetical science curriculum and a proposed opening course in biology. We want to ask two questions: What is the relevant domain of inquiry and what part of it should be taught in an introductory course? Biology is a broad discipline, with connections to physics and chemistry, and evolutionary biology is only one subset. Like its sister disciplines, evolutionary biology is not a system of beliefs, and this is precisely where science and religion part company. Religion is a system of beliefs within which it is entirely appropriate to ask "Do you believe in God?" and entirely appropriate to answer "Yes" without further discussion or summoning of evidence to support the response. It is not, however, appropriate to ask—as I am often asked—"Do you *really* believe in evolution?" where the "really" is stressed in order to emphasize the sheer ridiculousness of holding such a belief. This isn't a question that evolutionary biologists should answer. Ours is not a science based on belief, any more than physics, chemistry, psychology, or anthropology is based on belief. We gain confidence in a particular piece of evidence when it fits current theoretical predictions, when scientists from different laboratories can replicate the results, and when it fills a gap in our current understanding, leading in turn to new findings. A more sensible question, one worthy of a reply, is: "Do you *really* have confidence in the evidence of evolutionary biology?"

Note that this question restates the problem in two distinct ways, and we want students to appreciate the difference. We have shifted from "belief" to "confidence"—that is, from intuition and first-person experience to reliable methods of acquiring knowledge. And we have shifted from evolution as a process to evolutionary biology as a body of knowledge.

Part of an education in evolutionary biology includes—obligatorily, in my opinion—a discussion of current controversies, including what is contested and what is unknown. For example, there are heated discussions concerning the evolution of language, including debates over when this capacity emerged, the degree to which it is shared with other species, and whether it evolved for communication or for self-reflective thought and planning. Since the issue is by no means settled, and there are interesting theoretical predictions and relevant evidence for each of the different positions, this is a controversy worth airing in a textbook. When the advocates of intelligent design accuse evolutionary biologists of emulating ostriches by sticking their heads in the sand as a conflict-avoidance tactic, they have set up the ultimate straw man.

Sadly, several state legislatures have recently considered bills implying that our teachers avoid engaging in intellectual controversy. In Alabama, for example, bills in the House and the Senate proposed to protect the right of teachers "to present scientific critiques of prevailing scientific theories" and the right of students to "hold positions regarding scientific views"—as if this were not already the case. In Georgia, a House bill would have required teaching both "factual scientific evidence supporting or consistent with evolution theory and factual scientific evidence inconsistent with or not supporting the theory." I can't imagine a competent teacher of biology (or any other science) who would not engage in scientific critiques or who would not heartily support a student's interest in challenging what might appear to be dogma. But the question here is what counts as a legitimate challenge to current scientific accounts. Students at my own home institution often state that they "just don't believe" the evidence I present on the evolution of language or morality. My response is simple: Incredulity doesn't count as an alternative position or critique. I push students to banish incredulity as a line of attack and consider instead the nature of my assumptions: whether the results replicate, whether there are alternative interpretations that we might test with an observation or experiment. This is a discussion worth having with every student, and if they can learn the art, we will have done a great service to education and to human understanding more generally.

Many of the current controversies in evolutionary biology are not represented in high school textbooks because the alternative positions are insufficiently supported by empirical evidence; because the level of detail required to appreciate the debate is too great, given the (necessarily limited) scope of the course; or because the content of the controversy sits on the edges of neighboring disciplines, carrying the course too far afield. Unless publishers agree to make year-to-year updates of their textbooks, it would be prudent for the texts to stick closely not only to what is well known but what might be of interest down the line, including especially issues we can explore with the methods of science. I can think of several controversies not yet ready for textbook prime time because it is not yet clear whether they represent serious challenges to the dominant paradigms or small hiccups that will rapidly vanish due to insufficient evidence. For example, there are deep questions about the level, or target, of natural selection—questions that many evolutionary biologists debate today. Darwin largely thought about selection at the individual or, in some cases, group level. Group-level selection continued as a dominant theoretical perspective until the 1970s, and in many of the social sciences continues today. But the sociobiological revolution of the early 1970s caused a shift in focus from individuals and groups to genes. This revolution inspired thousands of studies, published in the best biology journals. In the 1980s, a relatively small group of evolutionary biologists began pushing once again for group selection, and they continue to do so. Although the group selection of today is different in some ways from past interpretations, with intriguing new evidence of its causal importance in human history, this is not yet a consensus view and so plays little or no role in high school biology. Which is as it should be—at least for now.

So here's a proposal. Let our science teachers do their job helping students understand the methods, theories, and discoveries of the sciences, including evolutionary biology. But let us also encourage our schools to facilitate the teaching of two other, complementary courses—or sections of a single course—forming a trio for each scientific discipline: one on the history and one on the interdisciplinary marriage between that particular science and the arts and humanities. I would love to see high school students take this kind of educational trio—an academic hat trick—focused on evolutionary biology. One section would introduce them to the methods, theory, and evidence in evolutionary biology, including discussions of natural selection, genetics, phylogeny, animal behavior, sexual reproduction, and speciation, to name a few. A second section would cover the history of evolutionary biology and include discussions of pre-Darwinian thought, Darwin, and such modern evolutionists as Fisher, Haldane, Hamilton, Williams, Trivers, Lewontin, Wilson, and Gould. It would also include controversies between biologists and religious scholars, starting with the debate in 1860 between Bishop Samuel Wilberforce and Thomas Henry

Huxley and culminating in today's debates between advocates and critics of the intelligent-design movement. A third section would look at how insights from evolutionary biology have led to exciting new discoveries in music, art, and morality, including similarities and differences across species in aesthetics and in perceptions of right and wrong. Students completing this hat trick would know how to distinguish science from other modes of inquiry and understanding, would appreciate the history of controversies within a given scientific discipline, and would understand how discoveries in the sciences often bleed over into neighboring areas, not taking them over but finding common ground while preserving methodological integrity.

There is perhaps no more important job than teaching. But our government must not force our teachers to blur the distinctions between science and other modes of inquiry, and must not blur what can be taught and what should be taught. Children exposed to an educational system that preserves the intellectual integrity of disciplines while building interdisciplinary bridges would be fortunate students indeed.

MARC D. HAUSER, an evolutionary psychologist, is a Harvard College Professor and a member of the Departments of Psychology, Organismic and Evolutionary Biology, and Biological Anthropology. He is co-director of the University's Mind, Brain, and Behavior Program and the author of *Wild Minds: What Animals Really Think* and the forthcoming *Moral Minds: The Unconscious Voice of Right and Wrong.*

SCOTT D. SAMPSON

Evoliteracy

It is all a question of story. We are in trouble just now because we do not have a good story. We are in between stories. The old story, the account of how we fit into it, is no longer effective. Yet we have not learned the new story.

—Thomas Berry,
The Dream of the Earth (1988)

DARWINIAN EVOLUTION forms the conceptual bedrock of the life sciences and has been fully accepted by the scientific community for well over a century. No less august an institution than the National Academy of Sciences has formally declared the theory of evolution to be "the central unifying concept of biology." Indeed, the renowned geneticist Theodosius Dobzhansky argued, "Nothing in biology makes sense except in the light of evolution."

If Dobzhansky's claim is true, we are in a lot of trouble. Several recent studies show that public understanding of evolution is woefully lacking, even among university science majors. In a November 2004 Gallup assessment of public opinion, only 35 percent of respondents agreed with the statement that "the theory of evolution is a scientific theory well supported by evidence." A few years earlier, another Gallup poll found 45 percent of the population in agreement that "God created humans pretty much in their present form at one time within the last 10,000 years." Other polls conducted over the past few years have revealed that about two-thirds of respondents favor the teaching of "intelligent design" along with evolution in science classrooms. To return to Dobzhansky's maxim: Given the abysmal level of public understanding of evolution, we are forced to the frightening conclusion that for the great majority of Americans, nothing in biology makes sense.

What lies behind this profound disconnect between scientists and the general public? The tendency in intellectual circles is to blame the evolution-education disaster on religious fundamentalists. There has been a strong anti-evolution faction in the United States since the initial publication of Darwin's *Origin of Species* almost 150 years ago. Creationists have often regarded a Darwinian world, in which humans are descended from ape-like primates and (worse yet) bacterial "slime," as a threat to the very meaning of life. Intelligent design, the most recent iteration of the creationist argument, has gained considerable support for its well-orchestrated "teach the controversy" campaign by appealing to Americans' sense of fairness. Yet the gross lack of public understanding of evolution cuts across the secular–religious spectrum, notably extending to university undergraduates—even biology majors who have taken courses in evolution. Moreover, many people who accept the truth of evolution mistakenly envision it as a completed process that has little bearing on human life, let alone on Earth's future biodiversity. So the efforts of anti-evolutionists, while certainly requiring a concerted response from the scientific community, do not represent the core of the crisis.

I consider the widespread failure to understand evolution to be a crisis of relevancy, intimately tied to the dominant mechanistic worldview. This reductionist perspective—a persistent legacy of sixteenth- and seventeenth-century scientists such as Bacon, Newton, Galileo, and Descartes—regards the natural world as a machine, to be understood through dissection of increasingly smaller parts. The Cartesian worldview, as it is often labeled, ultimately resulted in an explosion of specialized subdisciplines within science and a host of world-altering discoveries, from radioactivity to the germ theory of disease. Accompanying these insights, though, has been a strong bias toward viewing nature as something to be dominated rather than respected—a bias that has contributed directly to the current global sustainability crisis. Our relentless emphasis on the parts has left us with little conception, let alone understanding, of the whole. I would argue that our failure to communicate the essence of evolution is due largely to rampant compartmentalization within academia, which has left us with a void between two fundamental scientific concepts—ecology and evolution. Divorced from its ecological context and regarded as a process of the past, evolution has no appropriate meaning, at least for most people.

Ecologists and evolutionists tend to see the world through different eyes. Professional ecologists have long focused on processes operating on timescales too brief for evolution to be perceived. Evolutionary biologists, for their part, have been typically interested in lab experiments involving cells or genes, the effects of genetic change within populations, or the turnover of species through geologic time. There has been minimal cross-fertilization between these distinct research programs, and a synthesis of evolutionary and ecological theory has seemed unnecessary, or at least unapproachable. Hence university-educated science teachers have received little or no formal training in how these concepts are linked. Is it really surprising, then, that students rarely develop a deep understanding of them?

Fortunately, there is a movement afoot within both science and science education to bridge the eco-evolutionary gap, or at least to expand traditional divisions. Increasingly, scientists and granting agencies are seeking out interdisciplinary collaborations. Ecologists are beginning to look at regional and deep-time effects on ecosystems, while evolutionists are considering the role of ecosystem dynamics on evolutionary patterns and processes. Research on such topics as complex adaptive systems is uniting once disparate disciplines in a search for common explanations and even natural laws.

Simultaneously, radical new approaches to education are challenging our traditional notions of how learning should be accomplished. One small but growing example is the ecological literacy, or "ecoliteracy," movement. Advocates of ecoliteracy argue that designing entire curricula around key ecological concepts and outdoor activities will connect children with the natural world and foster growth of a more informed, more engaged citizenry. There are also several excellent resources now available, such as the UC Berkeley Understanding Evolution website (http://evolution.berkeley.edu), that address evolution education broadly. Yet all these efforts are just the beginning.

How might we make learning about evolution relevant? Some teachers of evolution have begun emphasizing the human benefits of understanding evolutionary concepts. For example, it is often noted that evolution has generated a tremendous diversity of organisms—primarily among plants and fungi—that together comprise an untapped (and rapidly disappearing) natural pharmacopeia, driving medical advances. But the array of human benefits derived from understanding evolution, while obviously relevant and important, is of limited educational utility, at least in the deeper sense. On its own, highlighting evolution in this way does little to establish the intimate and ancient connections that bond humans to the natural world. Indeed, like zoos, the human-benefits strategy provides implicit support for the prevailing, twisted notion that humans, the putative acme of evolution, are somehow entitled to dominate nature.

To be truly relevant, education must address the most pressing issues facing society. The global sustainability crisis, although not broadly appreciated (as evidenced by its virtual absence from the political landscape), is without doubt the most pressing issue of our time. Numerous authors have pointed out that any solution to this crisis must include reconnecting people with nature, so as to foster a passion that extends beyond our immediate social ties to embrace the natural world. Buddhists have long referred to this expression of loving-kindness toward nature as *metta*; more recently, famed biologist E. O. Wilson coined a new word for this phenomenon: "biophilia." A necessary prerequisite to *metta* or biophilia is a deep understanding of the world and our place within it. Such an understanding can be reached in several ways, including meditation and a lifetime spent in nature. However, for most urban dwellers—that is, for the primary consumers and polluters of Earth's resources—this major perceptual shift will have to involve comprehension of particular scientific concepts.

To my mind, of the many diverse concepts within science,

the two most in need of broad understanding are ecology and evolution. These revolutionary ideas—actually flipsides of the same coin—are the unifying themes of all the natural sciences. Ecology, the web of life, provides a description of the intricate, intertwined connections linking all organisms and their environments. Yet, while indispensable to the story of life, ecology alone is limited in its message. Evolution, largely overlooked by ecoliteracy advocates, provides essential content and meaning, as well as a missing dimension—time. Put simply, evolution is a description of change through time, and evolutionary studies investigate the processes that generated the wondrous, interconnected living world described by ecology. The study of evolution also encompasses changes in the physical realm, such as climate shifts and continental movements. Not surprisingly, it is highly interdisciplinary, including data from biology, geology, paleontology, geography, chemistry, physics, and ecology.

Viewed in this way, the web of life is composed of two distinct kinds of threads—those that link organisms at any given moment in time via the flow of energy (ecology), and those that link all life-forms throughout deep time via genetic information and shared ancestry (evolution). The fundamental messages of ecology are connections and interdependence, whereas for evolution they are change and interrelationship. Seen from this dual perspective, the two themes are complementary. Without an appreciation of evolution, our vision is myopic, effectively limited to the present day, and we cannot even begin to fathom the blossoming of life's diversity from single-celled ancestors. Without an appreciation of ecology, the intricate interconnections we share with the current panoply of life-forms cannot truly be envisioned. United in a single theme, the ecology-evolution duo provides a powerful lens through which to view life's web, forming the foundation of an integrated and underutilized perspective on nature.

I propose, then, that it is the links with ecology that make evolution relevant to our daily lives. As educators, we must demonstrate that the marvelous, interwoven complexity that characterizes every ecosystem, ancient and modern, is the result of a coevolutionary dance that has required millions upon millions of years. Thus, we require not only increased levels of ecoliteracy but also of "evoliteracy"—literacy in the workings of evolution such that it infuses one's life with certain kinds of meaning.

Evolution literacy, as put forth and defined here, entails comprehending such key concepts as inheritance, variation, natural selection, symbiosis, and extinction. The evoliterate person will understand that the evidence for life's shared ancestry is overwhelming, both in volume and observational detail. Equally important, evoliteracy, like its ecoliteracy counterpart, entails an understanding of place. What are the dominant plant communities that surround your home, and how long have they been there? Which of the trees are native to the region and which are exotic latecomers? What were some of the animals and plants that lived in your area before you did? What kinds of environments did these life-forms occupy, and what were some of the major physical events that generated those settings?

I live in Salt Lake City, Utah. Over the past 500 million years, my home has variously been open ocean, Saharan-style desert, giant lake, dense forest, and grassland. Swarms of ornamented, ocean-dwelling trilobites once grew, mated, and died here. Much later, dinosaurian behemoths strutted across balmy coastal plains beside a shallow inland seaway. Later still, woolly mammoths and saber-toothed cats sought forest shelter in an attempt to escape the icy winds descending from nearby glaciers. Evoliteracy, through an understanding of the past as well as the present, provides a deeper sense of place.

Equally important is the meaning that cascades from this new sense of place. An evoliterate individual will have grasped three profound evolutionary lessons:

1. *Earth time is deep time*. Humans suffer from severe temporal myopathy. Since our lifetimes are on the order of a century—not even an eyeblink in the geologic sense—1,000 years seems like an eternity. Thus, we have great difficulty even conceiving of a million years. Yet the roots of time run incredibly deep, with the age of the planet measured in billions of years. Never before in our hominid past has it been necessary to contemplate the remarkable depth of Earth's history, but that time has now arrived. Like the hour hand on a clock, our most pressing issues—global climate change, biodiversity losses, habitat destruction, and the like—occur at barely discernible rates. But when one truly begins to comprehend the immensity of time, it becomes increasingly difficult to maintain a sacred position for *Homo sapiens* alone atop the pedestal of life. Thus the transition from an anthropocentric to an ecocentric worldview will be aided by recognition of the true depth of Earth time.

2. *Change is the norm in natural systems*. Viewed from the perspective of deep time, our planet is in constant flux. Crustal plates migrate over the earth's surface, slamming into one another like a glacially paced and globally scaled version of bumper cars. Climates undergo major fluctuations, occasionally reaching hothouse or icehouse conditions, though remaining within the limits necessary for life. In response to tectonic activity and temperature variation, sea levels rise and fall dramatically. Life responds to these environmental assaults by means of evolution, which neces-

sarily includes the scythe of extinction. It is estimated that more than 99 percent of all species that have ever lived on Earth are now extinct. There have been five major extinctions, and we are in the midst of a sixth, perpetrated for the first time by a single species—us. Some biologists estimate that if current rates of human-induced extinctions continue, half of the 10 million to 30 million species with which we now share the biosphere will be extinguished by the close of this century. Paleontological evidence from previous mass-extinction events tells us that if the current ecological catastrophe is not stymied, eco-evolutionary recovery will require on the order of 5 million to 10 million years!

3. *All species, including humans, are inextricably interwoven into the history of life.* Life's unfolding is a tapestry in which every new thread is contingent on the nature, timing, and interweaving of the previous threads. If we could "rewind the tape" of life back to the beginning and let events play out again, the results would undoubtedly be dramatically different, even unrecognizable. Indeed, the odds of re-evolving primates (let alone humans) are virtually nil, even given billions of additional years. The origin of any new species is inextricably linked both to evolutionary history and to intricate ecological relationships with contemporary species. It is often said that the extinction of the dinosaurs 65 million years ago cleared the way for the radiation of mammals and ultimately the origin of humans. Yet the degree of life's interconnectedness far exceeds that implied in this statement. Dinosaurs persisted for 160 million years prior to this mass dying, coevolving in complex organic webs with plants, bacteria, fungi, and algae, as well as other animals. These Mesozoic life-forms mutually influenced not just one another's origins and fates but also those of all the species that followed. Had the major extinction of the dinosaurs occurred earlier

or later, or had dinosaurs never evolved, subsequent organisms would have been wholly different, and we would not be here to contemplate nature. An equivalent claim could be made for any major group at any point in the history of life. We owe our very existence to this unique, multibillion-year concatenation of events.

Since new species are generated from previously existing forms rather than appearing *de novo*, a corollary of this third lesson is that the history of life is a story of continual evolutionary jury-rigging rather than divine design. In this dynamic and contingent world, the law of the land (and the sea) is not so much survival of the fittest as survival of the fit enough. Natural selection is a non-random process, one that feeds on variation randomly generated with respect to organismal survival. One exciting hypothesis currently under consideration states that the evolution of organismal complexity has been driven largely by nature's unwavering propensity to break down energy gradients. Rather than removing meaning from life, an evolutionary perspective can and should fill us with a sense of wonder at the rich sequence of natural systems that gave us birth and continues to sustain us. Evoliteracy need not conflict with belief in a Creator; indeed, many evolutionists are deeply religious.

How, then, might we communicate a synthesis of ecology and evolution to a broad audience, given the complexity of both concepts? The answer will likely involve stories. Narrative has continually demonstrated its mettle in communicating science to a broad audience. More important, as noted in the quotation from the environmentalist Thomas Berry at the head of this essay, a new story is essential if we are to move toward a sustainable world. At this critical juncture in the history of our planet, what is the story that needs to be told? Returning to Berry's eloquence:

> Tell me the story of the river and the valley and the streams and woodlands and wetlands, of shellfish and finfish. A story of where we are and how we got here and the characters and roles that we play. Tell me a story, a story that will be my story as well as the story of everyone and everything about me, . . . a story that brings us together under the arc of the great blue sky in the day and the starry heavens at night. . . .

To my mind, ecology and evolution together provide a robust foundation for this story. Effective communication might take the form of multiple layered stories, or subplots. The layered-story approach allows educators to weave together several narratives and make non-intuitive connections about the workings of natural systems, as well as the role of humans within them. Manifestations of the eco-evolutionary tale could be told from the perspective of a single organism or an entire community; they could focus on a single modern habitat or a continent from the distant past. The narrative would include not only the flow of biological change but changes in the physical environment, from landscapes to climates. The strands of life's web extend beyond the biological to encompass the physical world, with continual feedback between the two realms, so these educational narratives must explore, and ultimately unite, Earth and life. Four story layers, or spheres, are particularly pertinent to unraveling any given ecosystem, past or present:

1. *Earth* (geosphere),
2. *Air* (atmosphere),
3. *Water* (hydrosphere), and
4. *Life* (biosphere).

Of course, ecology and evolution are most relevant to the biosphere (and of course this is where human history is also found) but their underlying themes, connection, and change apply equally to all four spheres. The degree of interconnectedness in natural systems is such that each of these story layers is incomplete without reference to the others; moreover, all four are animated and interconnected by the flow of energy and the cycling of matter, underscoring the need for some understanding of thermodynamics.

Several environmentalists, including Berry and Oberlin professor David Orr, have strongly advocated that rather than fragmenting the natural world into arbitrary disciplines, we should organize education around the categories suggested by nature itself—seashores, forests, rivers. Most important, we need to tell stories grounded in a sense of place. In keeping with the recent "constructivist" mandate for education, each version of the story could be tailored to the age and knowledge base of the audience, from children through university students. These stories need not and should not be presented as received truths. Meaning should be actively "constructed"—that is, students themselves should engage directly in the learning process, rather than acting as mere receptacles of information. The whole-to-parts approach advocated here is far more successful in generating meaning, and even transforming worldviews, than the traditional reductionist parts-to-whole method. Students could spend an entire school year exploring the four spheres as each relates to their particular town or region, with the goal of collectively writing their own story. Ideally, much of this learning would take place outdoors, in intimate contact with nature.

In sum, ecology and evolution together provide a robust scientific foundation for telling the big story—the story of who we are, how we got here, and our intimate links with nature. We must take up the challenge of unifying and demystifying these fundamental ideas.

There are many potential benefits of adopting a seamless eco-evolutionary approach to education. Perhaps most important, it would enable learners to make novel connections that help them to see themselves as fully embedded in the natural world. The World Wide Web and the global economy notwithstanding, the current worldview in so-called developed countries is based not on a deep understanding of connections and change but on their polar opposites—isolation and stasis. Due in large part to the Western reductionist mindset, we tend to see ourselves not as connected to the natural world but isolated from it, not as connected to communities in this country and abroad but isolated from them. Similarly, other than perhaps the progress of technology, we have great difficulty accepting change as a natural part of existence. This avoidance of change is reflected in anti-evolutionist views, along with the dangerous notion that Earth's resources are effectively limitless. With a perspective founded on insularity and equilibrium, rarely do we consider how our activities affect people elsewhere, let alone ecosystems or other species. Put simply, the natural systems in which we are embedded, and on which we are utterly dependent, are defined and organized by connections and change; our current worldview has simply not allowed us to see this reality.

The lengthy dominance of reductionism demonstrates that science can and does influence worldviews. We are now overdue for another paradigm shift in thought, and our current scientific understanding can once again help to provoke it. Connections and change, united, represent the foundation of a revised perspective that is both scientifically rigorous and culturally imperative. Notably, this view is also consistent with spiritual traditions like Buddhism and Hinduism, which have long regarded the nature of existence in terms of connections and change. Yet worldviews are not crafted overnight, or through a smattering of evolution teaching thinly spread over a dozen years of grade school. Therefore, our efforts should address not only the retention of evolution in science classrooms but its incorporation, along with ecology, into the very heart of the curriculum. We need a great deal more teaching of evolution, not less!

Here I strongly support calls from ecoliteracy advocates for systemic school reform, adding only that evolution must be incorporated as a sister concept to ecology. The eco-evolutionary message should be at the forefront in classrooms (and homes, natural history museums, botanical gardens, science centers, zoos, etc.) around the country. Imagine for a moment a world in which eco-evolutionary-centered education enabled children to appreciate and strengthen their intimate connections with nature both through the flow of energy in the living world and ancient kinship with all life-forms. Now imagine what kind of adults those children would become.

So rather than blaming religious fundamentalism for the failure of evolution education, we need to take ownership, as educators, of our central role in this persistent crisis. In particular, it is time to incorporate synthetic narratives that communicate the interrelated concepts of ecology and evolution. Evoliteracy is critical, and deserving of equal standing with ecoliteracy, because of evolution's fundamental role in biology and its unique power to convey essential concepts such as deep time, interrelationships, and change. Especially if this learning occurs in outdoor, natural laboratories, it will instill that all-important sense of wonder and passion for the nonhuman world that is currently in such short supply.

Such an education would present major obstacles for proponents of pseudoscience like intelligent design. Contrary to creationist arguments, an understanding of life's unfolding in terms of connections and change need not lead to an empty worldview based on sterile materialism. Indeed, I can envision nothing more beautiful than the dance of interconnectedness linking all life-forms on Earth across countless millennia. Ultimately, the integrated approach advocated here will foster the spread of biophilia and contribute to altering worldviews at a time when the sustainability of ecosystems, and perhaps our own species, is seriously imperiled.

Why is evoliteracy a crucial goal? Because evolution is central not only to the story of life as a whole but to other matters that affect our own lives daily—among them molecular biology, agriculture, biotechnology, conservation, even software development. Evolution-related issues currently include cloning, genetically modified foods, abuse of antibiotics, devastating species losses, pesticide use, and habitat destruction. Some will ask, "Why should we burden children (or adults, for that matter) with learning about the past, when there's so much to learn about the present?" The most compelling answer is that we live in a temporal snapshot, whereas Earth's systems operate on a vastly more immense timescale. We are here because of the deep-time interplay between ecology and evolution. Our continued survival and quality of life may well depend on broad public understanding of these concepts. Some knowledge of how things came to be is prerequisite to making wise decisions about the future. To paraphrase the philosopher Søren Kierkegaard, though life must be lived forward, it can only be understood backward. In a world careening out of control and plagued by such ills as global warming, overpopulation, habitat destruction, and rampant species loss, it is imperative that biology make sense.

SCOTT D. SAMPSON is a Canadian paleontologist who has a dual position at the University of Utah as chief curator at the Utah Museum of Natural History and associate professor in the Department of Geology and Geophysics. He recently served as the host of "Dinosaur Planet," a series of four animated television shows on the Discovery Channel.

Appendix

Excerpt from the Memorandum Opinion of The United States District Court for the Middle District of Pennsylvania, December 20, 2005[1]

Tammy Kitzmiller, et al. Case No. 04cv2688
Plaintiffs Judge John E. Jones III

v.

Dover Area School District, et al.
Defendants

4. WHETHER ID IS SCIENCE

After a searching review of the record and applicable case law, we find that while ID [intelligent design] arguments may be true, a proposition on which the Court takes no position, ID is not science. We find that ID fails on three different levels, any one of which is sufficient to preclude a determination that ID is science. They are: (1) ID violates the centuries-old ground rules of science by invoking and permitting supernatural causation; (2) the argument of irreducible complexity, central to ID, employs the same flawed

[1] **Editor's Note:** Judge Jones' case and trial-transcript citations have been omitted, for readability.

and illogical contrived dualism that doomed creation science in the 1980s; and (3) ID's negative attacks on evolution have been refuted by the scientific community. As we will discuss in more detail below, it is additionally important to note that ID has failed to gain acceptance in the scientific community, it has not generated peer-reviewed publications, nor has it been the subject of testing and research.

Expert testimony reveals that since the scientific revolution of the 16th and 17th centuries, science has been limited to the search for natural causes to explain natural phenomena. This revolution entailed the rejection of the appeal to authority, and by extension, revelation, in favor of empirical evidence. Since that time period, science has been a discipline in which testability, rather than any ecclesiastical authority or philosophical coherence, has been the measure of a scientific idea's worth. In deliberately omitting theological or "ultimate" explanations for the existence or characteristics of the natural world, science does not consider issues of "meaning" and "purpose" in the world. While supernatural explanations may be important and have merit, they are not part of science. This self-imposed convention of science, which limits inquiry to testable, natural explanations about the natural world, is referred to by philosophers as "methodological naturalism" and is sometimes known as the scientific method. Methodological naturalism is a "ground rule" of science today which requires scientists to seek explanations in the world around us based upon what we can observe, test, replicate, and verify.

As the National Academy of Sciences (hereinafter "NAS") was recognized by experts for both parties as the "most prestigious" scientific association in this country, we will accordingly cite its opinion where appropriate. NAS is in agreement that science is limited to empirical, observable, and ultimately testable data:

> Science is a particular way of knowing about the world. In science, explanations are restricted to those that can be inferred from the confirmable data—the results obtained through observations and experiments that can be substantiated by other scientists. Anything that can be observed or measured is amenable to scientific investigation. Explanations that cannot be based upon empirical evidence are not part of science.

This rigorous attachment to "natural" explanations is an essential attribute to science by definition and by convention. We are in agreement with Plaintiffs' lead expert Dr. [Kenneth R.] Miller, that from a practical perspective, attributing unsolved problems about nature to causes and forces that lie outside the natural world is a "science stopper." As Dr. Miller explained, once you attribute a cause to an untestable supernatural force, a proposition that cannot be disproven, there is no reason to continue seeking natural explanations, as we have our answer.

ID is predicated on supernatural causation, as we previously explained and as various expert testimony revealed. ID takes a natural phenomenon and, instead of accepting or seeking a natural explanation, argues that the explanation is supernatural.

Further support for the conclusion that ID is predicated on supernatural causation is found in the ID reference book to which ninth-grade biology students are directed [*Of Pandas and People*]. *Pandas* states, in pertinent part, as follows:

> Darwinists object to the view of intelligent design *because it does not give a natural cause explanation* of how the various forms of life started in the first place. Intelligent design means that various forms of life began abruptly, through an intelligent agency, with their distinctive features already

> intact—fish with fins and scales, birds with feathers, beaks, and wings, etc. (emphasis added).

Stated another way, ID posits that animals did not evolve naturally through evolutionary means but were created abruptly by a non-natural, or supernatural, designer. Defendants' own expert witnesses acknowledged this point.

It is notable that defense experts' own mission, which mirrors that of the IDM [intelligent design movement] itself, is to change the ground rules of science to allow supernatural causation of the natural world, which the Supreme Court in *Edwards* [*Edwards v. Arkansas*], and the court in *McLean* [*McLean v. Arkansas Board of Education*] correctly recognized as an inherently religious concept. First, defense expert Professor [Steve W.] Fuller agreed that ID aspires to "change the ground rules" of science and lead defense expert Professor [Michael] Behe admitted that his broadened definition of science, which encompasses ID, would also embrace astrology. Moreover, defense expert Professor [Scott] Minnich acknowledged that for ID to be considered science, the ground rules of science have to be broadened to allow consideration of supernatural forces.

Prominent IDM leaders are in agreement with the opinions expressed by defense expert witnesses that the ground rules of science must be changed for ID to take hold and prosper. William Dembski, for instance, an IDM leader, proclaims that science is ruled by methodological naturalism and argues that this rule must be overturned if ID is to prosper.

The Discovery Institute, the think tank promoting ID whose CRSC [Center for the Renewal of Science and Culture] developed the Wedge Document, acknowledges as "Governing Goals" to "defeat scientific materialism and its destructive moral, cultural and political legacies" and "replace materialistic explanations with the theistic understanding that nature and human beings are created by God." In addition, and as previously noted, the Wedge Document states in its "Five Year Strategic Plan Summary" that the IDM's goal is to replace science as currently practiced with "theistic and Christian science." The IDM accordingly seeks nothing less than a complete scientific revolution in which ID will supplant evolutionary theory.[2] Notably, every major scientific association that has taken a position on the issue of whether ID is science has concluded that ID is not, and cannot be considered as such.

Initially, we note that NAS, the "most prestigious" scientific association in this country, views ID as follows:

---

[2] Further support for this proposition is found in the Wedge Strategy, which is composed of three phases: Phase I is scientific research, writing and publicity; Phase II is publicity and opinion-making; and Phase III is cultural confrontation and renewal. In the "Five Year Strategic Plan Summary," the Wedge Document explains that the social consequences of materialism have been "devastating" and that it is necessary to broaden the wedge with a positive scientific alternative to materialistic scientific theories, which has come to be called the theory of ID. "Design theory promises to reverse the stifling dominance of the materialist worldview, and to replace it with a science consonant with Christian and theistic convictions." Phase I of the Wedge Strategy is an essential component and directly references "scientific revolutions." Phase II explains that alongside a focus on influential opinion-makers, "we also seek to build up a popular base of support among our natural constituency, namely, Christians. We will do this primarily through apologetics seminars. We intend these to encourage and equip believers with new scientific evidence that support the faith, as well as to 'popularize' our ideas in the broader culture." Finally, Phase III includes pursuing possible legal assistance "in response to resistance to the integration of design theory into public school science curricula."

> Creationism, intelligent design, and other claims of supernatural intervention in the origin of life or of species are not science because they are not testable by the methods of science. These claims subordinate observed data to statements based on authority, revelation, or religious belief. Documentation offered in support of these claims is typically limited to the special publications of their advocates. These publications do not offer hypotheses subject to change in light of new data, new interpretations, or demonstration of error. This contrasts with science, where any hypothesis or theory always remains subject to the possibility of rejection or modification in the light of new knowledge.

Additionally, the American Association for the Advancement of Science (hereinafter "AAAS"), the largest organization of scientists in this country, has taken a similar position on ID, namely, that it "has not proposed a scientific means of testing its claims" and that "the lack of scientific warrant for so-called 'intelligent design theory' makes it improper to include as part of science education. . . ." Not a single expert witness over the course of the six-week trial identified one major scientific association, society, or organization that endorsed ID as science. What is more, defense experts concede that ID is not a theory as that term is defined by the NAS and admit that ID is at best "fringe science" which has achieved no acceptance in the scientific community.

It is therefore readily apparent to the Court that ID fails to meet the essential ground rules that limit science to testable, natural explanations. Science cannot be defined differently for Dover students than it is defined in the scientific community as an affirmative action program, as advocated by Professor Fuller, for a view that has been unable to gain a foothold within the scientific establishment. Although ID's failure to meet the ground rules of science is sufficient for the Court to conclude that it is not science, out of an abundance of caution and in the exercise of completeness, we will analyze additional arguments advanced regarding the concepts of ID and science.

ID is at bottom premised upon a false dichotomy, namely, that to the extent evolutionary theory is discredited, ID is confirmed. This argument is not brought to this Court anew, and in fact, the same argument, termed "contrived dualism" in *McLean*, was employed by creationists in the 1980s to support "creation science." The court in *McLean* noted the "fallacious pedagogy of the two-model approach" and that "[i]n efforts to establish 'evidence' in support of creation science, the defendants relied upon the same false premise as the two-model approach . . . all evidence which criticized evolutionary theory was proof in support of creation science." We do not find this false dichotomy any more availing to justify ID today than it was to justify creation science two decades ago.

ID proponents primarily argue for design through negative arguments against evolution, as illustrated by Professor Behe's argument that "irreducibly complex" systems cannot be produced through Darwinian, or any natural, mechanisms. However, we believe that arguments against evolution are not arguments for design. Expert testimony revealed that just because scientists cannot explain today how biological systems evolved does not mean that they cannot, and will not, be able to explain them tomorrow.

As Dr. [Kevin] Padian aptly noted, "absence of evidence is not evidence of absence." To that end, expert testimony from Drs. Miller and Padian provided multiple examples where *Pandas* asserted that no natural explanations exist, and in some cases that none could exist, and yet natural explanations have been identified in the intervening years. It also bears mentioning that as Dr. Miller stated, just because scientists cannot explain every evolutionary detail does not undermine its validity as a scientific theory as no theory in science is fully understood.

As referenced, the concept of irreducible complexity is ID's alleged scientific centerpiece. Irreducible complexity is a negative argument against evolution, not proof of design, a point conceded by defense expert Professor Minnich. Irreducible complexity additionally fails to make a positive scientific case for ID, as will be elaborated upon below.

We initially note that irreducible complexity as defined by Professor Behe in his book *Darwin's Black Box* and subsequently modified in his 2001 article entitled "Reply to My Critics," appears as follows:

> By irreducibly complex I mean a single system which is composed of several well-matched, interacting parts that contribute to the basic function, wherein the removal of any one of the parts causes the system to effectively cease functioning. An irreducibly complex system cannot be produced directly by slight, successive modifications of a precursor system, because any precursor to an irreducibly complex system that is missing a part is by definition nonfunctional. . . . Since natural selection can only choose systems that are already working, then if a biological system cannot be produced gradually it would have to arise as an integrated unit, in one fell swoop, for natural selection to have anything to act on.

Professor Behe admitted in "Reply to My Critics" that there was a defect in his view of irreducible complexity because, while it purports to be a challenge to natural selection, it does not actually address "the task facing natural selection." Professor Behe specifically explained that "[t]he current definition puts the focus on removing a part from an already functioning system," but "[t]he difficult task facing Darwinian evolution, however, would not be to remove parts from sophisticated pre-existing systems; it would be to bring together components to make a new system in the first place." In that article, Professor Behe wrote that he hoped to "repair this defect in future work"; however, he has failed to do so even four years after elucidating his defect.

In addition to Professor Behe's admitted failure to properly address the very phenomenon that irreducible complexity purports to place at issue, natural selection, Drs. Miller and Padian testified that Professor Behe's concept of irreducible complexity depends on ignoring ways in which evolution is known to occur. Although Professor Behe is adamant in his definition of irreducible complexity when he says a precursor "missing a part is by definition nonfunctional," what he obviously means is that it will not function in the same way the system functions when all the parts are present. For example, in the case of the bacterial flagellum, removal of a part may prevent it from acting as a rotary motor. However, Professor Behe excludes, by definition, the possibility that a precursor to the bacterial flagellum functioned not as a rotary motor but in some other way, for example as a secretory system.

As expert testimony revealed, the qualification on what is meant by "irreducible complexity" renders it meaningless as a criticism of evolution. In fact, the theory of evolution proffers exaptation as a well-recognized, well-documented explanation for how systems with multiple parts could have evolved through natural means. Exaptation means that some precursor of the subject system had a different, selectable function before experiencing the change or addition that resulted in the subject system with its present function. For instance, Dr. Padian identified the evolution of the mammalian middle-ear bones from what had been jawbones as an example of this process. By defining irreducible complexity in the way that he has, Professor Behe attempts to exclude the phenomenon of exaptation by definitional fiat, ignoring, as he does so, abundant evidence which refutes his argument.

Notably, the NAS has rejected Professor Behe's claim for irreducible complexity by using the following cogent reasoning:

> [S]tructures and processes that are claimed to be "irreducibly" complex typically are not on closer inspection. For example, it is incorrect to assume that a complex structure or biochemical process can function only if all its components are present and functioning as we see them today. Complex biochemical systems can be built up from simpler systems through natural selection. Thus, the "history" of a protein can be traced through simpler organisms. . . . The evolution of complex molecular systems can occur in several ways. Natural selection can bring together parts of a system for one function at one time and then, at a later time, recombine those parts with other systems of components to produce a system that has a different function. Genes can be duplicated, altered, and then amplified through natural selection. The complex biochemical cascade resulting in blood clotting has been explained in this fashion.

As irreducible complexity is only a negative argument against evolution, it is refutable and accordingly testable, unlike ID, by showing that there are intermediate structures with selectable functions that could have evolved into the allegedly irreducibly complex systems. Importantly, however, the fact that the negative argument of irreducible complexity is testable does not make testable the argument for ID. Professor Behe has applied the concept of irreducible complexity to only a few select systems: (1) the bacterial flagellum; (2) the blood-clotting cascade; and (3) the immune system. Contrary to Professor Behe's assertions with respect to these few biochemical systems among the myriad existing in nature, however, Dr. Miller presented evidence, based upon peer-reviewed studies, that they are not in fact irreducibly complex.

First, with regard to the bacterial flagellum, Dr. Miller pointed to peer-reviewed studies that identified a possible precursor to the bacterial flagellum, a subsystem that was fully functional, namely the Type-III secretory system. Moreover, defense expert Professor Minnich admitted that there is serious scientific research on the question of whether the bacterial flagellum evolved into the Type-III secretory system, the Type-III secretory system into the bacterial flagellum, or whether they both evolved from a common ancestor. None of this research or thinking involves ID. In fact, Professor Minnich testified about his research as follows: "[W]e're looking at the function of these systems and how they could have been derived one from the other. And it's a legitimate scientific inquiry."

Second, with regard to the blood-clotting cascade, Dr. Miller demonstrated that the alleged irreducible complexity of the blood-clotting cascade has been disproven by peer-reviewed studies dating back to 1969, which show that dolphins' and whales' blood clots despite missing a part of the cascade, a study that was confirmed by molecular testing in 1998. Additionally and more recently, scientists published studies showing that in puffer fish, blood clots despite the cascade missing not only one but three parts. Accordingly, scientists in peer-reviewed publications have refuted Professor Behe's predication about the alleged irreducible complexity of the blood-clotting cascade. Moreover, cross-examination revealed that Professor Behe's redefinition of the blood-clotting system was likely designed to avoid peer-reviewed scientific evidence that falsifies his argument, as it was not a scientifically warranted redefinition.

The immune system is the third system to which Professor Behe has applied the definition of irreducible complexity. Although in *Darwin's Black Box*, Professor Behe wrote that not only were there no natural explanations for the immune system at the time but that natural explanations were impossible regarding its origin. However, Dr. Miller presented peer-reviewed studies refuting Professor Behe's claim that the immune system was irreducibly complex. Between 1996 and 2002, various studies confirmed each element of the evolutionary hypothesis explaining the origin of the immune system. In fact, on cross-examination, Professor Behe was questioned concerning his 1996 claim that science would never find an evolutionary explanation for the immune system. He was presented with fifty-eight peer-reviewed publications, nine books, and several immunology textbook chapters about the evolution of the immune system; however, he simply insisted that this was still not sufficient evidence of evolution, and that it was not "good enough."

We find that such evidence demonstrates that the ID argument is dependent upon setting a scientifically unreasonable burden of proof for the theory of evolution. As a further example, the test for ID proposed by both Professors Behe and Minnich is to grow the bacterial flagellum in the laboratory; however, no one inside or outside of the IDM, including those who propose the test, has conducted it. Professor Behe conceded that the proposed test could not approximate real-world conditions, and even if it could, Professor Minnich admitted that it would merely be a test of evolution, not design.

We therefore find that Professor Behe's claim for irreducible complexity has been refuted in peer-reviewed research papers and has been rejected by the scientific community at large. Additionally, even if irreducible complexity had not been rejected, it still does not support ID, as it is merely a test for evolution, not design.

We will now consider the purportedly "positive argument" for design encompassed in the phrase used numerous times by Professors Behe and Minnich throughout their expert testimony, which is the "purposeful arrangement of parts." Professor Behe summarized the argument as follows: We infer design when we see parts that appear to be arranged for a purpose. The strength of the inference is quantitative; the more parts that are arranged, the more intricately they interact, the stronger is our confidence in design. The appearance of design in aspects of biology is overwhelming. Since nothing other than an intelligent cause has been demonstrated to be able to yield such a strong appearance of design, Darwinian claims notwithstanding, the conclusion that the design seen in life is real design is rationally justified. As previously indicated, this argument is merely a restatement of the Reverend William Paley's argument applied at the cell level. Minnich, Behe, and Paley reach the same conclusion, that complex organisms must have been designed using the same reasoning, except that Professors Behe and Minnich refuse to identify the designer, whereas Paley inferred from the presence of design that it was God. Expert testimony revealed that this inductive argument is not scientific and as admitted by Professor Behe, can never be ruled out.

Indeed, the assertion that design of biological systems can be inferred from the "purposeful arrangement of parts" is based upon an analogy to human design. Because we are able to recognize design of artifacts and objects, according to Professor Behe, that same reasoning can be employed to determine biological design. Professor Behe testified that the strength of the analogy depends upon the degree of similarity entailed in the two propositions; however, if this is the test, ID completely fails.

Unlike biological systems, human artifacts do not live and reproduce over time. They are non-replicable, they do not undergo genetic recombination, and they are not driven by natural selection. For human artifacts, we know the designer's identity, human, and the mechanism of design, as we have experience based upon empirical evidence that humans can make such things, as well as many other attributes including the designer's abilities, needs, and desires. With ID, proponents assert that they refuse to propose hypotheses on the designer's identity, do not propose a mechanism, and the designer, he/she/it/they, has never been seen. In that vein, defense expert Professor Minnich agreed that in the case of human artifacts and objects, we know the identity and capacities of the human designer but we do not know any of those attributes for the designer of biological life. In addition, Professor Behe agreed that for the design of human artifacts, we know the designer and its attributes and we have a baseline for human design that does not exist for design of biological systems. Professor Behe's only response to these seemingly insurmountable points of disanalogy was that the inference still works in science fiction movies.

It is readily apparent to the Court that the only attribute of design that biological systems appear to share with human artifacts is their complex appearance, i.e. if it looks complex or designed, it must have been designed. This inference to design based upon the appearance of a "purposeful arrangement of parts" is a completely subjective proposition, determined in the eye of each beholder and his/her viewpoint concerning the complexity of a system. Although both Professors Behe and Minnich assert that there is a quantitative aspect to the inference, on cross-examination they admitted that there is no quantitative criteria for determining the degree of complexity or number of parts that bespeak design, rather than a natural process. As Plaintiffs aptly submit to the Court, throughout the entire trial only one piece of evidence generated by Defendants addressed the strength of the ID inference: The argument is less plausible to those for whom God's existence is in question and is much less plausible for those who deny God's existence.

Accordingly, the purported positive argument for ID does not satisfy the ground rules of science which require testable hypotheses based upon natural explanations. ID is reliant upon forces acting outside of the natural world, forces that we cannot see, replicate, control, or test, which have produced changes in this world. While we take no position on whether such forces exist, they are simply not testable by scientific means and therefore cannot qualify as part of the scientific process or as a scientific theory.

It is appropriate at this juncture to address ID's claims against evolution. ID proponents support their assertion that evolutionary theory cannot account for life's complexity by pointing to real gaps in scientific knowledge, which indisputably exist in all scientific theories, but also by misrepresenting well-established scientific propositions.

Before discussing Defendants' claims about evolution, we initially note that an overwhelming number of scientists, as reflected by every scientific association that has spoken on the matter, have rejected the ID proponents' challenge to evolution. Moreover, Plaintiffs' expert in biology, Dr. Miller, a widely recognized biology professor at Brown University who has written university-level and high school biology textbooks used prominently throughout the nation, provided unrebutted testimony that evolution, including common descent and natural selection, is "overwhelmingly accepted" by the scientific community and that every major scientific association agrees. As the court in *Selman* [*Selman et al. v. Cobb County School District and Board of Education*] explained, "evolution is more than a *theory* of origin in the context of science. To the contrary, evolution is the dominant *scientific* theory of origin accepted by the majority of scientists" (emphasis in original). Despite the scientific community's overwhelming support for evolution, Defendants and ID proponents insist that evolution is unsupported by empirical evidence. Plaintiffs' science experts, Drs. Miller and Padian, clearly explained how ID proponents generally and *Pandas* specifically distort and misrepresent scientific knowledge in making their anti-evolution argument.

In analyzing such distortion, we turn again to *Pandas*, the book to which students are expressly referred in the disclaimer. Defendants hold out *Pandas* as representative of ID and Plaintiffs' experts agree in that regard. A series of arguments against evolutionary theory found in *Pandas* involve paleontology, which studies the life of the past and the fossil record. Plaintiffs' expert Professor Padian was the only testifying expert witness with any expertise in paleontology.[3] His testimony therefore remains unrebutted. Dr. Padian's demonstrative slides, prepared on the basis of peer-reviewing scientific literature, illustrate how *Pandas* systematically distorts and misrepresents established, important evolutionary principles.

We will provide several representative examples of this distortion. First, *Pandas* misrepresents the "dominant form of understanding relationships" between organisms, namely, the tree of life, represented by classification determined via the method of cladistics. Second, *Pandas* misrepresents "homology," the "central concept of comparative biology," that allowed scientists to evaluate comparable parts among organisms for classification purposes for hundreds of years. Third, *Pandas* fails to address the well-established biological concept of exaptation, which involves a structure changing function, such as fish fins evolving fingers and bones to become legs for weight-bearing land animals. Dr. Padian testified that ID proponents fail to address exaptation because they deny that organisms change function, which is a view necessary to support abrupt appearance. Finally, Dr. Padian's unrebutted testimony demonstrates that *Pandas* distorts and misrepresents evidence in the fossil record about pre-Cambrian-era fossils, the evolution of fish to amphibians, the evolution of small carnivorous dinosaurs into birds, the evolution of the mammalian middle ear, and the evolution of whales from land animals.

[3] Moreover, the Court has been presented with no evidence that either Defendants' testifying experts or any other ID proponents, including *Pandas'* authors, have such paleontology expertise as we have been presented with no evidence that they have published peer-reviewed literature or presented such information at scientific conferences on paleontology or the fossil record.

In addition to Dr. Padian, Dr. Miller also testified that *Pandas* presents discredited science. Dr. Miller testified that *Pandas'* treatment of biochemical similarities between organisms is "inaccurate and downright false" and explained how *Pandas* misrepresents basic molecular biology concepts to advance design theory through a series of demonstrative slides. Consider, for example, that he testified as to how *Pandas* misinforms readers on the standard evolutionary relationships between different types of animals, a distortion which Professor Behe, a "critical reviewer" of *Pandas* who wrote a section within the book, affirmed.[4] In addition, Dr. Miller refuted *Pandas'* claim that evolution cannot account for new genetic information and pointed to more than three dozen peer-reviewed scientific publications showing the origin of new genetic information by evolutionary processes. In summary, Dr. Miller testified that *Pandas* misrepresents molecular biology and genetic principles, as well as the current state of scientific knowledge in those areas in order to teach readers that common descent and natural selection are not scientifically sound.

Accordingly, the one textbook to which the Dover ID Policy directs students contains outdated concepts and badly flawed science, as recognized by even the defense experts in this case.

A final indicator of how ID has failed to demonstrate scientific warrant is the complete absence of peer-reviewed publications supporting the theory. Expert testimony

---

[4] Additionally, testimony provided by Professor Behe revealed an increasing gap between his portrayal of ID theory and how it is presented in *Pandas*. Although he is a "critical reviewer" of the work, he disagrees with language provided in the text, including but not limited to the text's very definition of ID.

revealed that the peer-review process is "exquisitely important" in the scientific process. It is a way for scientists to write up their empirical research and to share the work with fellow experts in the field, opening up the hypotheses to study, testing, and criticism. In fact, defense expert Professor Behe recognizes the importance of the peer-review process and has written that science must "publish or perish." Peer review helps to ensure that research papers are scientifically accurate, meet the standards of the scientific method, and are relevant to other scientists in the field. Moreover, peer review involves scientists submitting a manuscript to a scientific journal in the field, journal editors soliciting critical reviews from other experts in the field and deciding whether the scientist has followed proper research procedures, employed up-to-date methods, considered and cited relevant literature and generally, whether the researcher has employed sound science.

The evidence presented in this case demonstrates that ID is not supported by any peer-reviewed research, data, or publications. Both Drs. Padian and [Barbara] Forrest testified that recent literature reviews of scientific and medical-electronic databases disclosed no studies supporting a biological concept of ID. On cross-examination, Professor Behe admitted that: "There are no peer reviewed articles by anyone advocating for intelligent design supported by pertinent experiments or calculations which provide detailed rigorous accounts of how intelligent design of any biological system occurred." Additionally, Professor Behe conceded that there are no peer-reviewed papers supporting his claims that complex molecular systems, like the bacterial flagellum, the blood-clotting cascade, and the immune system, were intelligently designed. In that regard, there are no peer-reviewed articles supporting Professor Behe's argument that certain complex molecular structures are "irreducibly complex."[5] In addition to failing to produce papers in peer-reviewed journals, ID also features no scientific research or testing.

After this searching and careful review of ID as espoused by its proponents, as elaborated upon in submissions to the Court, and as scrutinized over a six-week trial, we find that ID is not science and cannot be adjudged a valid, accepted scientific theory, as it has failed to publish in peer-reviewed journals, engage in research and testing, and gain acceptance in the scientific community. ID, as noted, is grounded in theology, not science. Accepting for the sake of argument its proponents' as well as Defendants' argument that to introduce ID to students will encourage critical thinking, it still has utterly no place in a science curriculum. Moreover, ID's backers have sought to avoid the scientific scrutiny, which we have now determined that it cannot withstand, by advocating that the *controversy* but not ID itself should be taught in science class. This tactic is at best disingenuous, and at worst a canard. The goal of the IDM is not to encourage critical thought but to foment a revolution which would supplant evolutionary theory with ID.

To conclude and reiterate, we express no opinion on the ultimate veracity of ID as a supernatural explanation.

---

[5] The one article referenced by both Professors Behe and Minnich as supporting ID is an article written by Behe and Snoke entitled "Simulating evolution by gene duplication of protein features that require multiple amino acid residues." A review of the article indicates that it does not mention either irreducible complexity or ID. In fact, Professor Behe admitted that the study which forms the basis for the article did not rule out many known evolutionary mechanisms and that the research actually might support evolutionary pathways if a biologically realistic population size were used.

However, we commend to the attention of those who are inclined to superficially consider ID to be a true "scientific" alternative to evolution without a true understanding of the concept the foregoing detailed analysis. It is our view that a reasonable, objective observer would, after reviewing both the voluminous record in this case and our narrative, reach the inescapable conclusion that ID is an interesting theological argument but that it is not science.

H. CONCLUSION

The proper application of both the endorsement and Lemon tests[6] to the facts of this case makes it abundantly clear that the [Dover School] Board's ID Policy violates the Establishment Clause. In making this determination, we have addressed the seminal question of whether ID is science. We have concluded that it is not, and moreover that ID cannot uncouple itself from its creationist, and thus religious, antecedents.

Both Defendants and many of the leading proponents of ID make a bedrock assumption which is utterly false. Their presupposition is that evolutionary theory is antithetical to a belief in the existence of a supreme being and to religion in general. Repeatedly in this trial, Plaintiffs' scientific

---

[6] **Editor's Note:** According to the Pew Forum on Religion and Public Life, the Lemon test "is a three-part test . . . frequently used to determine whether a government action violates the Establishment Clause. Under this test, an action must (1) have a bona fide secular purpose; (2) not advance or inhibit religion; and (3) not excessively entangle the government with religion. If the challenged action fails any of the three parts of the Lemon test, it is deemed to have violated the Establishment Clause."

experts testified that the theory of evolution represents good science, is overwhelmingly accepted by the scientific community, and that it in no way conflicts with, nor does it deny, the existence of a divine creator. To be sure, Darwin's theory of evolution is imperfect. However, the fact that a scientific theory cannot yet render an explanation on every point should not be used as a pretext to thrust an untestable alternative hypothesis grounded in religion into the science classroom or to misrepresent well-established scientific propositions.

The citizens of the Dover area were poorly served by the members of the Board who voted for the ID Policy. It is ironic that several of these individuals, who so staunchly and proudly touted their religious convictions in public, would time and again lie to cover their tracks and disguise the real purpose behind the ID Policy.

With that said, we do not question that many of the leading advocates of ID have bona fide and deeply held beliefs which drive their scholarly endeavors. Nor do we controvert that ID should continue to be studied, debated, and discussed. As stated, our conclusion today is that it is unconstitutional to teach ID as an alternative to evolution in a public-school science classroom. Those who disagree with our holding will likely mark it as the product of an activist judge. If so, they will have erred, as this is manifestly not an activist Court. Rather, this case came to us as the result of the activism of an ill-informed faction on a school board, aided by a national public-interest law firm eager to find a constitutional test case on ID, who in combination drove the Board to adopt an imprudent and ultimately unconstitutional policy. The breathtaking inanity of the Board's decision is evident when considered against the factual backdrop which has now been fully revealed through this trial. The students, parents, and teachers of the Dover Area School District deserved better than to be dragged into this legal maelstrom, with its resulting utter waste of monetary and personal resources. To preserve the separation of church and state mandated by the Establishment Clause of the First Amendment to the United States Constitution, and Art. I, § 3 of the Pennsylvania Constitution, we will enter an order permanently enjoining Defendants from maintaining the ID Policy in any school within the Dover Area School District, from requiring teachers to denigrate or disparage the scientific theory of evolution, and from requiring teachers to refer to a religious, alternative theory known as ID. We will also issue a declaratory judgment that Plaintiffs' rights under the Constitutions of the United States and the Commonwealth of Pennsylvania have been violated by Defendants' actions. Defendants' actions in violation of Plaintiffs' civil rights as guaranteed to them by the Constitution of the United States and 42 U.S.C. § 1983 subject Defendants to liability with respect to injunctive and declaratory relief, but also for nominal damages and the reasonable value of Plaintiffs' attorneys' services and costs incurred in vindicating Plaintiffs' constitutional rights.

NOW, THEREFORE, IT IS ORDERED THAT:

1. A declaratory judgment is hereby issued in favor of Plaintiffs pursuant to 28 U.S.C. §§ 2201, 2202, and 42 U.S.C. § 1983 such that Defendants' ID Policy violates the Establishment Clause of the First Amendment of the Constitution of the United States and Art. I, § 3 of the Constitution of the Commonwealth of Pennsylvania.

2. Pursuant to Fed.R.Civ.P. 65, Defendants are permanently enjoined from maintaining the ID Policy in any school within the Dover Area School District.
3. Because Plaintiffs seek nominal damages, Plaintiffs shall file with the Court and serve on Defendants, their claim for damages and a verified statement of any fees and/or costs to which they claim entitlement. Defendants shall have the right to object to any such fees and costs to the extent provided in the applicable statutes and court rules.

s/ John E. Jones III
United States District Judge

ALSO EDITED BY JOHN BROCKMAN

CURIOUS MINDS

How a Child Becomes a Scientist

John Brockman has collected the essays of twenty-seven scientists who have written about what is was that sent them on the path of their life's work. For Robert Sapolsky, Stanford professor of biology, it was an argument with a rabbi over a passage in the Bible. Physicist Lee Smolin traces his inspiration to a volume of Einstein's work, picked up as a diversion from heartbreak. Mihaly Csikszentmihalyi, a psychologist and the author of *Flow*, found his calling through Descartes. In essays by Murray Gell-Mann, Nicholas Humphrey, and Freeman J. Dyson, illuminating memoir meets superb science writing in stories that invite us to consider what it is—and what it isn't—that sets the scientific mind apart.

Biography/Science/1-4000-7686-2

THE NEXT FIFTY YEARS

Science in the First Half of the Twenty-first Century

The Next Fifty Years collects twenty-five never-before-published essays by the world's most visionary scientists. Theoretical physicist Paul Davies examines the likelihood that by the year 2050 there will be a human presence on Mars; psychiatrist Nancy Etcoff explains current research into the creation of emotion-sensing jewelry that could tell us when to take an antidepressant pill; and evolutionary biologist Richard Dawkins explores the probability that we will soon be able to obtain a genome printout that predicts our natural end. This fascinating book explores not only the practical possibilities of the near future but also the social and political ramifications of the developments of the strange new world to come.

Science/0-375-71342-5